Learning Through School Science Investigation

Learning Through School Science by student

Azra Moeed · Dayle Anderson

Learning Through School Science Investigation

Teachers Putting Research into Practice

 Springer

Azra Moeed
School of Education
Victoria University of Wellington
Wellington, New Zealand

Dayle Anderson
School of Education
Victoria University of Wellington
Wellington, New Zealand

ISBN 978-981-13-4655-2   ISBN 978-981-13-1616-6   (eBook)
https://doi.org/10.1007/978-981-13-1616-6

This Springer imprint is published by the registered company Springer Nature Singapore Pte Ltd.
The registered company address is: 152 Beach Road, #21-01/04 Gateway East, Singapore 189721,
Singapore

Foreword

Why is this Book Needed?

Practical work has a central role in science, and in science education. Science is driven by curiosity. The aim of science is to find explanations for the phenomena and events that we see happening around us. What are the constituents of the material world, what can they do or have done to them and how can we explain their behaviour? Science seeks to provide explanations that are convincing and persuasive, based on evidence and careful reasoning, and open to critique in order to test, refine and improve them.

Practical work is also a prominent, distinctive and essential feature of science education. Like science itself, learning science is also driven by curiosity. We get curious about something we observe, but often need to look at it more closely to try to see exactly what is happening. Sometimes, we have an idea about why something happens, but need to test it by seeing if our predictions are borne out in practice. So practical work that involves observing, handling and manipulating real objects and materials is an essential component of any effective science education.

The role of practical work in science education, however, as the authors of this book acknowledge, is different from its role in science itself. The aim of scientific research is to add to public knowledge—to find out things that no one previously knew or understood. The aim of science education, on the other hand, is to help students develop their personal understanding of the ideas and models that scientists use to explain its behaviour, and of the methods of enquiry and forms of reasoning that underpin convincing knowledge claims. Simply 'telling' students is unlikely to work, and leads to dull and uninspiring teaching. Effective science education involves 'showing' learners certain things, or putting them into situations where they can see things for themselves. Practical experiences matter.

How, then, do we square this with the findings of many research studies that, whilst students invariably say that they like and enjoy practical work in their science classes, there is little evidence that it leads to significant gains in understanding of scientific concepts, ideas or explanations? Students may later recall what they did and observed in a practical science activity, but often cannot say what question the activity was trying to answer, what it showed, or how the observations might be explained. The appropriate response to this, as this book demonstrates, is not to question *whether* practical work should be such a prominent part of school science education, but to think harder about *how* it is used in the science classroom. Although set within the current context of science education in New Zealand, the issues it explores are recognised in many, indeed most, countries. And the strategies it adopts for developing and improving practice are also ones that could be applied anywhere.

Essentially, the authors' approach is not to offer teachers ready-made practical activities or materials but rather to encourage them to reflect on particular aspects of their current practice. The research team supported their collaborating teachers in collecting focused evidence on how they currently use practical work, reflecting on this within a collegial framework involving teachers and researchers, setting specific personal targets for change, implementing these and monitoring the outcomes.

A great strength of this book is the detailed accounts it provides of practical activities implemented by a group of experienced primary and secondary teachers committed to the use of practical work in their teaching. The accounts, supported by photographs and examples of student work, bring these lessons to life and give a real sense of the classroom atmosphere. The figures and chapter appendices that show some of the tools the researchers used to focus and stimulate teachers' reflection on specific aspects of their practice are illuminating and useful. Too often in educational research, case studies provide insufficient information on the context and setting of the work being reported. Not here. We learn enough about these teachers and their professional settings to make informed judgments about the applicability of their learning and insights to other settings.

Several of the teachers comment on the importance of connecting practical activity to students' ideas and interests. One primary teacher says that a practical activity 'has to be … something interesting and something that they want to find out about.' A secondary teacher argues that 'for students, the most important thing is to find what *they* want to find out and do what *they* want to do'. Ownership of the question is clearly important. Research in cognitive science shows, however, that open and unstructured inquiry is rarely if ever an effective learning strategy. Students may choose to explore questions that are trivial or lead nowhere very useful. The challenge for teachers is to guide students to pose good questions—ones that the students themselves understand and value—which can also lead towards useful insights and knowledge.

The key messages for practice that the authors draw out from their work include the importance of being clear about the intended learning outcomes of any given practical activity—and not having too many of them for any single activity. They also highlight the importance of thinking about what would count as evidence that learning outcomes had been achieved, and of making explicit to students the ways in which their actions in successfully undertaking a practical task illustrate and exemplify scientific practices and values. They conclude that 'when teachers are explicit in modelling, identifying and describing scientific behaviours, students begin to adopt them for themselves and associate them with science'.

These are valuable insights, directly applicable to the science curricula of many countries. They can help science teachers and researchers to design and implement practical work in ways that more effectively achieve the conceptual, procedural and epistemic goals of school science education.

York, UK

Prof. Robin Millar
OBE, Department of Education
University of York

Preface

As science teacher educators, we have always felt very strongly about the importance of supporting our student teachers to engage their students in purposeful practical science investigation, also known in some parts of the world as science inquiry. Our experience as teachers has shown us the power of science investigation for drawing students into science, seeing the joy of developing and satisfying curiosity (although it sometimes creates more questions than it answers!), and for helping students consider how science itself works—the kinds of questions that it can and can't answer, what is valued and the qualities and kinds of thinking that are involved. However, we have noticed that many student teachers, at least initially, simply think of investigating in science as the 'fun' part of science, rather than connecting it to purposeful learning. This attitude reflects our own research concerning school students' views of practical science—they too mostly see it as fun, but not particularly useful for learning.

Evidence internationally suggests that students' experiences of science investigation are often limited. Furthermore, where students do undertake some practical science investigation in schools, there is little evidence that they learn much by doing it. So, are we correct as science teacher educators to place so much emphasis on science investigation, or should we put our energies elsewhere? We have consequently been thinking carefully about the role and purpose of science investigation in school science. How do teachers and their students at different levels of the education system perceive science investigation and view their role in learning? Just what do students learn through science investigation? What kind of learning is best achieved through science investigation and how? We are not the first researchers interested in such questions and many recommendations have been made in the extant literature. An important focus for us was to find out what happens when teachers change their practice in ways that are informed by research. What impact does such a change have on student experience and learning?

This book is an outcome of a research project that set out to find answers to these questions.[1] Initially, we investigated teachers' views and beliefs, and their existing classroom practice. We explored the complex transactions between teacher and learner that occur before, during and after investigations and how they influence opportunities to learn science. We noted that many of the practices being used by teachers involved in the project were effective in supporting student learning. In the early findings chapters, using rich descriptions of teacher beliefs and practices together with student work and voice, we show what teachers were thinking and doing and what students were learning. Our analysis of this practice was shared with the teachers along with current research recommendations. Teachers decided on aspects of their practice that they wanted to change to enhance student learning. We researched these changes in practice and in the later chapters we share how they impacted on student learning and perceptions of science.

The aims of the book are several. First, it aims to share insights from observations of teachers' practice of science investigation, highlighting beliefs and practices that support effective learning. Second, it describes what happens when teachers use research literature and evidence from their teaching and student learning to enhance their practice of science investigation in their programmes. We discuss successful changes and ongoing challenges in teachers' practice of science investigation. Finally, it is a book about classroom research—it describes methods and tools that may be useful to others interested in learning from and improving science teaching practice.

The teachers whose practices and change stories are described here include a teacher of 5-year-olds new to school, a teacher of 9–10-year-olds in their middle primary/elementary years and three teachers of 13–15-year-olds teaching a compulsory general science programme in the same junior department of a secondary/high school. Although the study is set in New Zealand, we anticipate the findings and issues discussed are applicable in all countries that aim to provide their students with rich experiences of science that build not only their understanding of the knowledge produced by science, but also of the way in which science itself functions and its relevance to their everyday lives.

In our experience teachers are learners who wish to use their learning to enhance the outcomes of their students, and so this book would be applicable to anyone interested in teaching science. It will be particularly useful to:

- those involved in science teacher education
- those interested in science education research
- science teachers and pre-service science teachers
- teachers who want to focus on inquiry into science teaching and learning as personal or departmental professional development.

---

[1] This research was funded by Ministry of Education Teaching and Learning Research Initiative.
  Moeed, A., Anderson, D., Rofe, C., & Bartholomew, R. (2016). Beyond play: Learning through science investigation (Report submitted to the New Zealand Council for Educational Research, New Zealand). Retrieved from http://www.tlri.org.nz/tlri-research/research-completed/school-sector/beyond-play-learning-through-science-investigation.

Chapter 1 begins with some relevant terminology and their definitions. We then examine the nature, role and purposes of school science investigation, current evidence on student learning from science investigation and its implications for teaching practice. Chapter 2 describes the New Zealand context and curriculum requirements and the research design and methodology. Chapters 3 and 4 respectively present the participating primary and secondary teachers' existing practices that support student learning through science investigation. Chapters 5 and 6 describe the changes the teachers made and their impact on student learning. Finally, in Chap. 7, we reflect on our learnings from this research and discuss the implications for curriculum, policy and practice.

Wellington, New Zealand   Azra Moeed
                                                                                 Dayle Anderson

Acknowledgements

We would like to thank The Teaching and Learning Initiative for funding and the teachers and students who participated in the research which made this book possible. We are appreciative of the time given to us by Prof. Robin Millar as our advisor, and to Assoc. Prof. Miles Barker for his critique and advice. We thank Susan Kaiser for editorial support.

Contents

1 **Introduction: School Science Investigation—What Research
  Says?** 1
  1.1 School Science Investigation: What Research Says 1
  1.2 What Do We Mean by Science Investigation? 1
  1.3 Science Investigation and School Science Investigation 3
  1.4 Why Teach Science? What Are the Goals of Science
      Education? 6
  1.5 Multiple Purposes for Learning Through Science Investigation ... 9
  1.6 Learning from School Science Investigation 12
  1.7 Summary 13
  References 14

2 **The New Zealand Context and Research Design** 17
  2.1 Science in the New Zealand Curriculum 17
  2.2 Research Design and Methodology 21
  2.3 Summary 25
  References 30

3 **Science Investigation in Primary School** 33
  3.1 The NZ Primary Context and Curriculum 33
  3.2 Participants and Contexts 34
  3.3 Teachers' Beliefs About Science Investigation 36
  3.4 Teachers' Practice of Science Investigation 38
  3.5 Teacher Strategies that Provided Opportunities for Learning
      Through Investigation 50
  3.6 Student Learning from Investigation 57
  3.7 Summary 59
  References 68

4 Science Investigation in Secondary School 71
   4.1 Teacher Beliefs About Nature of Science Investigation 72
   4.2 Teacher Practices that Support Student Investigation 74
   4.3 Student Learning from Investigation 83
   4.4 Summary 89
   References 91

5 Science Investigation in Primary School: Changes to Teacher
   Practice 93
   5.1 Research-Informed Reflection 93
   5.2 Primary Teachers' Intended Changes 94
   5.3 Observed Changes to Primary Teachers' Practice 95
   5.4 Impact of Changes of Primary Students' Learning 101
   5.5 Summary 107
   References 108

6 Science Investigation in Secondary School: Changes to Teacher
   Practice 109
   6.1 Research-Informed Reflection 109
   6.2 Changes the Teachers Intended to Make 110
   6.3 Observed Changes to Secondary Teachers' Practice 111
   6.4 Summary 124
   References 130

7 Enhancing Learning Through School Science Investigation 131
   7.1 Teacher Beliefs About Science Investigation 131
   7.2 Approaches to Science Investigation 132
   7.3 Teacher Practices that Support Substantive and Syntactic
        Understanding 133
   7.4 The Influence of Change in Practice on Student Learning
        Through Science Investigation 135
   7.5 Teacher Agency 136
   7.6 Science Investigations: Further Opportunities for Learning 137
   7.7 Suggestions for Practice 139
   References 139

Chapter 1
Introduction: School Science Investigation—What Research Says?

1.1 School Science Investigation: What Research Says

It seems logical to include science investigation in school science education programmes, after all, it is through practical investigation that scientists have developed scientific theories and ideas about how the world works. However, what counts as science investigation? Are all hands-on practical activities investigation? How does school science investigation reflect investigation carried out by scientists? Why should science investigation be part of school science? What do we hope students will learn? And importantly, what is our current understanding of what students actually learn from investigation in school science?

In this chapter, we review the recent literature relevant to these questions. We begin by defining some of the terms commonly used in this context and compare the similarities and differences between investigations carried out by scientists and those carried out in school. Then, in order to understand more clearly the role of science investigation, we discuss the goals of science education itself. Finally, we examine the literature concerning student learning from science investigation.

1.2 What Do We Mean by Science Investigation?

Even in the usually precise world of research literature, different terminologies are prevalent. In literature emanating from the United States, science investigation is more commonly referred to as 'science inquiry'. Inquiry can in itself be a confusing term. In many countries, inquiry learning often refers to a generic teaching approach focused on students developing and answering their own questions. While this approach is readily applicable to science and student investigation, it is also used to apply to a range of disciplines and is often cross- or multidisciplinary in nature. To add to the confusion, in school science teachers and students use the words

© Springer Nature Singapore Pte Ltd. 2018
A. Moeed and D. Anderson, *Learning through School Science Investigation*,
https://doi.org/10.1007/978-981-13-1616-6_1

practical, experiment, practical work and investigation interchangeably. The term science investigation is often used to refer to any laboratory work, fair testing, inquiry, as well as teacher or student-led investigation. In this book, we use the following definitions in relation to school science:

> **Practical Work**
> *Practical work is any science teaching and learning activity where students observe or manipulate objects individually or in small groups. Practical work has generally been called hands-on activities or doing science; it may or may not take place in the laboratory.* (Millar, 2004)

> **Experiments**
> *An experiment is a specific form of investigation where an intervention is performed to create a phenomenon that can be observed either quantitatively (by measuring) or qualitatively. Experiments are different to an investigation where data are collected by observation or what is happening already (Hackling, 1993). The key criterion is '... whether you have to do something to create a phenomenon you are interested in and hence generate data, or can simply observe what is happening, or what exists, anyhow.'* (Millar, 2015a, p. 418)

> **School Science Investigation**
> *School science investigation is an activity requiring identification of a question, using both conceptual and procedural knowledge in planning and carrying out the investigation, gathering, processing and interpreting data and drawing conclusions based on evidence. Ideally, the process is iterative and the student has some choice in what they want to investigate.* (Millar, 2010a, p. 108)

(For further detail, see Abrahams & Millar, 2008; Moeed, 2015.)
Experiments may be carried out within an investigation or by themselves. Practical work can be inclusive of experiments, practical activities and investigations. In this book, our focus is on school science investigation as defined above.

1.3   Science Investigation and School Science Investigation

Science investigation has a single purpose—to generate new knowledge. The term *substantive* is used to describe the body of knowledge generated by a discipline, which in science is concerned with understanding the physical, material and natural world (Schwab, 1964). Different science disciplines tend to have their own approaches to investigation and scientists develop rigorous procedures and protocols to follow when investigating any science discipline in order to ensure the generation of reliable and valid evidence. Scientific theories are developed by scientists to try to explain a body of evidence and can be revised as new evidence comes to light or evidence is reconsidered in the light of an alternative explanation. Before scientific claims and theories are accepted by the scientific community, the robustness of the investigation design, the validity and reliability of the evidence are subject to peer critique. The adequacy of explanations and theories is also open to critique based on the evidence. Ford and Forman (2006) suggest that scientists have the dual role of 'Constructors and Critiquers' of scientific claims (p. 15). Scientists are always aware of and seek peer review and critique at every stage of their investigation. Understanding these processes of knowledge creation and acceptance is called *syntactic* knowledge (Schwab, 1964).

Ideally, school science investigation is in many ways similar to that carried out by scientists (see Fig. 1.1). It is an activity in which students solve a problem or answer a question about their physical or natural surroundings. Students use their substantive and syntactic knowledge to plan and carry out an investigation, collecting, processing and interpreting data, and generating explanations based on evidence. The process of investigation is not linear or sequential. Students should reflect on their planning and critically evaluate the decisions they have made and the evidence they have collected, and thoughtfully consider the explanations they are offering. However, there are critical differences. As scientists are working at the cutting edge of their discipline, their goal is to create new knowledge and challenge existing theories, whereas students are often learning about established substantive science concepts. They are also developing their knowledge about how science is done, their syntactic knowledge. Scientists are experts in their fields and even when investigating in multidisciplinary teams, they have a depth of knowledge and understanding of procedures. Importantly, they often work within one specialist field and can combine their substantial substantive understanding with their procedural knowledge in designing their methodology. Students, on the other hand, have some knowledge but instead of working within a specific field are learning a variety of investigative approaches and procedures in a broad range of scientific domains.

When school curricula suggest that students will experience a variety of approaches to science investigation, the intention is for students to learn that scientists investigate in different ways, depending on their discipline. Schwab (1964) states that scientists' investigations are guided by the substantive structures that are partially tied to a phenomenon in a particular discipline. A physicist working with

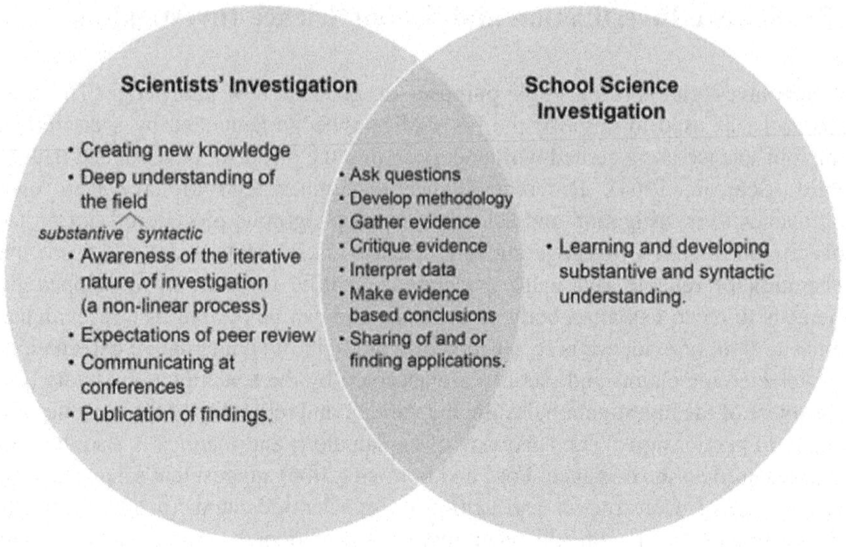

Scientists' Investigation

- Creating new knowledge
- Deep understanding of
  the field

substantive syntactic

- Awareness of the iterative
  nature of investigation
  (a non-linear process)
- Expectations of peer review
- Communicating at
  conferences
- Publication of findings.

- Ask questions
- Develop methodology
- Gather evidence
- Critique evidence
- Interpret data
- Make evidence
  based conclusions
- Sharing of and or
  finding applications.

**School Science
Investigation**

- Learning and developing
  substantive and syntactic
  understanding.

Fig. 1.1 Similarities and differences between scientists' science investigation and school science investigation

high-energy particles and a biologist concerned with the flow of energy in an ecosystem are interested in vastly different kinds of phenomena. So, they bring to their investigation different substantive structures and different ways of viewing the phenomenon which leads to different approaches they take to investigate. In school science, for this reason, investigation in biology should reflect the nature of that discipline, the same applying for chemistry and physics (Schwab, 1962). Yet, we are expecting students to learn to investigate in a range of science disciplines, including physics, chemistry, biology, geology and astronomy, of which, contrary to investigations carried out by scientists, they have very little substantive or syntactic understanding, perhaps even none.

School science investigations have been classified in a number of ways. For example, Schwab (1962) used the notion of 'degrees of freedom' students are given, ranging from following given instructions to deciding on a question and making choices about the procedures to follow by themselves. Investigating in science (Ministry of Education, 1995) classifies them as open investigations, closed investigations, and complete investigations. Open investigations provide opportunities for students to develop attitudes and skills, for example, being persistent, being systematic in collecting data and data analysis, describing trends and offering explanations. The teacher's role in open investigations is monitoring, guiding, and challenging students to work out the most appropriate procedures. Roberts (2009) states:

> Genuine open-ended investigations… are those in which pupils are unaware of any correct
> answer, where there are many different routes to a valid solution, where choices have to be

made about equipment selection, where different sources of uncertainty lead to variation in the data and where students reflect and modify their practice in the light of the evidence they have collected. The evidence produced, then, is messy rather than the laundered version common in practical work contrived to illustrate ideas to students. (p. 31)

Such an investigation, she argues, allows students to be creative in their problem-solving. In a class where such creativity is allowed, no two investigations need be the same as all students would have licence to come up with their own approach. In contrast, closed investigations have more defined outcomes and students make few decisions in either planning or carrying out the investigation. These are typically 'recipe practicals' where students follow a set of given instructions to arrive at an often known conclusion. Closed investigations can include teacher demonstrations.

Hodson (2014) proposes four phases when students are conducting a science investigation (inquiry):

(1) A design and planning phase, during which specific research questions are asked and goals clarified, hypotheses formulated (if appropriate), investigative procedures devised and data collection techniques selected.
(2) A performance phase, during which the various operations are carried out and data are collected.
(3) A reflection phase, during which findings are considered and interpreted in relation to various theoretical perspectives, conclusions drawn and justifications for those conclusions formulated and refined.
(4) A recording and reporting phase, during which the procedure, its rationale and the various findings, interpretations and conclusions are recorded for personal use and expressed in the style approved by the community for communication to, and critical scrutiny by, others. (p. 9)

At any age, students can use their skills of identifying and classifying, data gathering, processing and interpreting, and reporting to carry out a complete investigation. Importantly, in complete investigations students have some freedom to decide what they wish to investigate and have an appropriate level of independence. Teachers have an important role in developing student confidence and providing opportunities for students to be reflective and critical, and over time develop an understanding of the nature of science and science investigation. The above classification is based on the degree of control and ownership students have in deciding on the question, planning and carrying out the investigation. As the students are learning to investigate, age- and stage-appropriate guidance is needed from the teacher. School science investigation may form a continuum from being teacher-led to student-led.

Opportunities for student ownership provide openings for learning in science that other forms of practical work do not. However, comparing school science investigation with those of scientists helps us to see the complexity of the learning demands involved. In any one investigation, there are multiple new experiences and opportunities for students to learn. They could be grappling with a substantive idea, or learning to use a different approach to investigation, or identifying what it means to 'be' scientific, or, and most likely, a combination of all three! At the same time, they are also learning to manage themselves, work with others, and make decisions while considering a range of options. What does this complexity mean for the way in which we implement science investigations in our classrooms to maximise learning? To help us consider this question, we first examine the goals of science education itself.

1.4   Why Teach Science? What Are the Goals of Science Education?

What do we hope to achieve through science education, and how do these goals affect what and how we teach? The purposes of science education in a given curriculum or context impact on why we include science investigation in school science and the learning we hope to achieve from it. Science education is seen as a vital subject in most international curricula as science offers important explanations about the material world. It helps students understand the world that they live in and how it works. Living in a world that is increasingly reliant on science to answer or solve the many problems we face, there is a need for students to participate and succeed in education and to carry on in science beyond their schooling. Equally important is that school science education produces citizens who can make informed decisions about the socio-scientific issues that they face in their everyday lives. These needs form two distinct goals for science education.

Historically, science curricula have been framed largely by scientists who have considered school science as preparation for university entrance and future careers in science. The perceived societal need for future scientists was the priority for producing a workforce that provided for a modern lifestyle. There is currently some debate regarding how students engage with science at school in order to prepare for each of these goals. Are the science learning needs of those studying science for the purpose of becoming informed citizens who can use the products of science and can make informed choices about socio-scientific issues they face in their everyday lives the same as those studying with a career in science in mind? Traditionally, preparing future scientists, engineers and other professionals have had a narrower, discipline-bound focus (Tytler, 2007). Stuckey, Hofstein, Mamlok-Naaman, and Eilks (2013) argue that preparing learners for further science studies requires a structured and discipline-based curriculum strictly aligned with the guided acquisition of knowledge. Fensham (2004) criticises the so-called 'pipeline', a pre-professional education that delivers science-ready students to colleges and universities. Such an approach for preparing future scientists is considered outdated by many who argue that the science learning needs of future scientists are no different to the needs of those preparing for citizenship. Tytler (2007) argues for the need for a reimagined curriculum which is diverse beyond the traditional focus on canonical science ideas. He proposes that school science should foster capabilities such as reasoning, creativity, communication, and problem-solving, and include an understanding about the nature of science and how it interacts with society.

One shortcoming of a narrow discipline-based curriculum is that it serves the needs of less than 10% of the students and some claim that such a curriculum does not suit the majority of students who were unlikely to continue with science education past the compulsory years (Millar, 2011; Osborne & Dillon, 2008; Tytler & Osborne, 2012). Fensham (2004) argues that like most dissatisfied consumers or customers, those students whose needs are not addressed by such a curriculum no longer wish to continue with science, perhaps because the course content is not perceived to be relevant. Student disengagement with science remains an issue internationally and led to Tytler and Osborne (2012) investigating students' attitude towards science and their aspirations for science learning. Students' disinterest in science is being expressed earlier in their education than previously observed (Bolstad & Hipkins, 2008; Tytler, 2007).

Concerns over not only student but public disenchantment with, and ignorance of science has led to a commonly adopted goal of developing scientific literacy in the general population, the intention being for citizens to have the tools to critically evaluate what science has to offer and to make informed choices. The current answer to the 'why' question about science education in many countries is now for *all* to become scientifically literate and for *some* to carry on with science for utilitarian, economic and developmental reasons (Bull, Gilbert, Barwick, Hipkins, & Baker, 2010; Lederman, 2007; Millar & Osborne, 1998).

Modern curricula commonly require students to learn about established and accumulated scientific knowledge that is appropriate for their age, interests and needs; additionally, they are expected to develop an understanding of the processes by which scientific knowledge is produced and tested to provide the basis for our confidence in that knowledge (Millar, 2004). Again, we see complexity in what ought to be taught and learnt, given the vast and growing amount of scientific knowledge and the many different methods scientists use for generating the knowledge within science that could potentially be learnt. The curriculum requirements reflect both kinds of disciplinary knowledge—substantive and syntactic.

There is a consensus amongst science education researchers that developing syntactic understanding about the nature of science is essential for scientific literacy for citizenship (e.g. Abd-El-Khalick, 2012; Allchin, 2014; Hodson & Wong, 2014; Lederman & Lederman, 2014). Most believe that scientific knowledge is tentative, empirical, inferential, subjective, creative, requires theory building and is socially and culturally located (Abd-El-Khalick, 2012; Anderson & Moeed, 2017; Lederman, 2004; Lederman & Lederman, 2014). Most modern science curricula require teachers to teach students about the nature of science, and that students ought to learn it. However, there is less agreement on which ideas about science are crucial to include in the modern school science curricula and how these nature of science ideas ought to be taught and learnt (Osborne, Collins, Ratcliffe, Millar, & Duschl, 2003).

Briefly, Lederman and Lederman (2004) argue for a generalised list of features of scientific knowledge. However, concerns have been raised about the over generalisation and presentation of these ideas or tenets as they may lead to a surface approach rather than to a deeper understanding. For example, suggesting that scientific knowledge is tentative may lead to students believing that it is unreliable. Hodson and Wong (2014) argue that scientific knowledge is reliable, citing examples of science ideas taught in school. Christodoulou and Osborne (2014) assert that whilst scientific theories are 'always open to revision, [their] core elements are stable and beyond reasonable doubt' (p. 1275).

Hodson and Wong (2017) propose a more contextualised approach to syntactic learning stating that students best develop an understanding about the nature of science through working alongside scientists. Their view has merit but is perhaps not workable in mass education. Osborne (2017) agrees with Hodson and Wong's (2017) critique of the consensus view about the tenets of the nature of science but he is critical of Hodson and Wong's alternative, arguing that it does not provide sufficient detail or conceptual clarity. Allchin (2011, 2014, 2017) suggests that students need to develop a functional understanding of the nature of science through engaging in contextually relevant and recent case studies of real science problems. Allchin (2013, 2017) further proposes a 'whole science' approach: 'The context of the NOS education determines the scope of the epistemic knowledge required' (p. 20). Allchin (2017) argues that compared to the consensus view, the list of nature of science elements that students should understand is extensive, inclusive and open-ended. He suggests that if we focus on a core consensus list, students would not be fully prepared for functional scientific literacy. In his view, students need to have ways to learn how to deal with nature of science issues as they arise. To do this, they have to be habitually asking 'How do we know this?' as well as 'Based on this evidence and expertise, how can we have trust in this particular scientific claim?' (p. 20). His suggestion for developing such habits is through epistemic analysis of the nature of science with structured authentic cases. Thus, science education could prepare students with skills and ways of thinking that enable citizens to continue learning about the nature of science after leaving school (Allchin, 2013).

This active goal for learning how science works has been expressed recently in the Framework for K-12 Standards (National Research Council, 2012). These standards emphasise teaching about investigations as *scientific practices*. Osborne (2014) argues that it is a better representation of current understanding of the nature of science as a social and cultural practice. Christodoulou and Osborne (2014) argue that:

> The process of empirical inquiry cannot exist in isolation from the theories that it seeks to test, the analysis and interpretation of the data, and the arguments required to resolve conflicting interpretations. Moreover, such a model of science demands that students be taught explicitly not only domain-specific content knowledge but a body of basic procedural and epistemic knowledge that is essential to engage in the critical evaluation of any scientific activity or report. (p. 1275)

These practices include: asking questions; developing and using models; constructing explanations; engaging in argument from evidence; planning and carrying out investigation; analysing and interpreting data; using mathematical and computational thinking; and obtaining, evaluating and communicating information (Osborne, 2014). It could be argued that participating in school science investigation provides many opportunities for such practices.

Taking on board this development in thinking about the goals of science education, it is now generally acknowledged that it is not just enough to understand content; if the goal is for all students to be able to engage effectively with science as citizens, they also need to understand how scientific knowledge is created and validated, and have the skills to critique scientific claims, as well as apply the knowledge they have in making decisions in their everyday lives. How can science investigation support these multiple learning demands?

1.5 Multiple Purposes for Learning Through Science Investigation

When science curricula require the learning of both substantive and syntactic content, the burden can fall heavily on school science investigation. Engaging in a school science investigation has the potential for learning that addresses the goals described above. It can illustrate science ideas and help develop substantive learning; it can support students to develop investigative skills; it can provide opportunities for critique of methods and claims, and to consider the nature of scientific knowledge and how it becomes accepted.

We can, therefore, identify three major purposes for school science investigation (see Fig. 1.2).

Teachers commonly incorporate investigations as a means of delivering on the first of these purposes—to develop substantive science ideas. For example, through exploration, teachers may help students to construct their understanding about electric currents. Students can investigate what happens when a number of bulbs are wired in a series or in parallel. They may observe that as the number of bulbs (of the same voltage) is increased in series, the bulbs become dimmer. When wired in parallel, adding another bulb does not affect the brightness of the ones already in the circuit. The teacher may consider this to be a useful illustration, and there is evidence which suggests that experiencing a scientific phenomenon is memorable for students. When allowed to experiment by themselves some students say they find it easier to remember: 'When I do it myself, and the result is in front of my eyes, I can remember it' (Moeed, 2010, p. 148, see also Toplis, 2012). However, the explanation as to why an electric current behaves the way it does is not evident from the observation. The leap from observations of behaviour to scientific theories that explain a range of gathered evidence is a big one (Millar, 2004, 2010a, 2010b). As Osborne (2014) observed, students need experience at developing and critiquing

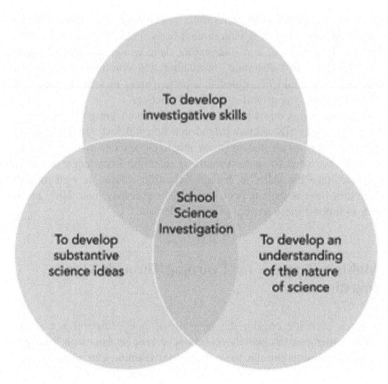

Fig. 1.2 Purposes of school science investigation

possible explanations and models rather than accepting one given model or explanation.

Understanding electrical currents is typical of much science where the explanatory ideas are complex and not easily derived from observations (Millar, 2004). Helping students understand the scientific explanation of currents requires careful sequencing of several learning experiences that provide some evidence of the accepted theory and also the opportunity to consider other theories. It can be said that *thinking* behind the *doing* is needed to develop an understanding as to why currents behave in this way. If the purpose of investigation is for students to understand the scientific explanation and its epistemology, it needs to be supported by a sequence of activities and carefully managed discussion (Mortimer & Scott, 2003).

While there is a degree of freedom in the above activity, the students have often not come up with a question to investigate and often will have little input into the design. A common scenario is that they have followed the recipe provided by the teacher to arrive at a conclusion the teacher wanted them to reach and verify. This type of closed investigation can give students the impression that science investigations are linear and the result is already known. Literature suggests that such recipe practicals are common in New Zealand secondary schools and elsewhere

(Hipkins et al., 2002). While they can be productive in illustrating and developing a substantive idea, they provide limited examples of investigation. Students need to have other opportunities to come to appreciate the complexity of the nature of science investigation.

A second purpose for wiring electrical circuits in series and parallel could be to develop procedural skills. Students may often be given circuit diagrams and instructed how to connect the components. They may well learn how to wire the circuits together in which case one could say they have learnt some procedural skills. Procedural skills in themselves, while necessary, are not sufficient if the goal is to learn ways to investigate in science.

Developing students' investigative skills are often viewed by teachers as the need for students to learn 'the scientific method.' This term describes investigating in science as a linear process of defining the problem, developing an appropriate hypothesis, gathering reliable data to test the hypothesis by planning, carrying out experiments or observations, and drawing a conclusion in support or otherwise of the tested hypothesis (Moeed, 2013; Windschitl, Thompson, & Braaten, 2007; Wong, Hodson, Kwan, & Yung, 2008). Many science educators and researchers are critical of the use of the term 'scientific method'. They consider following a series of ordered steps to be an inadequate description of scientific practice. There are many ways to investigate and science investigations often do not follow such a linear and sequential process (Windschitl et al., 2007). Asking students to identify and try out various ways of connecting and arranging circuit components may provide more opportunities to learn to investigate scientifically, i.e. develop investigative rather than procedural skills. Such an open investigation provides greater scope for considering the reliability of evidence and offering evidence-based conclusions (Glaesser, Gott, Roberts, & Cooper, 2009).

The third of these purposes, addressing the nature of science, is possibly the biggest task. Teachers often assume that simply by carrying out science investigations students will learn about the empirical nature of investigation and about the nature of science. In the substantive and procedural examples above, there are opportunities to develop nature of science understandings. For example, in investigating what happens when a number of bulbs are wired in series or in parallel, students would have opportunities to consider the empirical nature of science. In investigating the effect of different arrangements of components in the circuit students may be scaffolded to identify and critique evidence. However, while possibilities for learning are apparent, research suggests that such learning may be problematic. The evidence is, perhaps unsurprisingly, that students do not readily make connections between what they are doing in school science and the nature of science as a discipline (Driver, Leach, Millar, & Scott, 1996; Lederman, 2004; Lederman & Lederman, 2014). We, therefore, turn now to examining the literature in relation to student learning through science investigation.

1.6 Learning from School Science Investigation

Current evidence suggests that while students undertake some practical, hands-on science investigation in schools, there is little evidence that they learn much by doing it (Anderson, 2012; Hume & Coll, 2008; Millar, 2010a, 2010b; Moeed, 2010). The research on what students learn from science investigation is limited. Secondary school teachers report that investigative work is motivational, and students see it as an enjoyable alternative to written work (Moeed, 2010; Goodrum, Rennie, & Hackling, 2001). According to Toplis (2012), students in the UK see practical work as an important component of their school science experience because: it is active and provides interest; there is an opportunity to work with peers; it gives them control and it is preferable to writing and rote learning.

Similar experiences prevail in primary schools, but often little is done in the classroom to connect investigation with either substantive or epistemological learning and students thus tend to see practical work as fun but disconnected from their science learning. These observations support the view that younger students naively see science theories as directly observable (Driver et al., 1996; Lederman, 2007). In a case where primary students were presented with both syntactic and substantive learning opportunities through science investigation, they more readily recognised substantive learning outcomes. These students appeared to view substantive ideas as the natural outcome of science education (Anderson, 2012).

Moeed (2010) found that when asked about the science they had learnt, secondary school students almost always talked about the science concepts; for example, when asked what they had learnt about energy they talked about gravitational potential energy changing into kinetic energy. At times, when they did bring up learning about science investigation, they were more focused on the process they had to follow for assessment of investigation rather than the purpose of doing the science investigation. There was little mention that would indicate the development of procedural knowledge or understanding of the very nature of science investigation. Learning from school science investigation seems to depend on what students anticipate as being the expected learning in science. These findings connect with others who suggest that if students are to develop syntactic understanding, then such learning needs to be made explicit (e.g. Abd El-Khalick & Lederman, 2000).

Roberts (2009) give primacy to students developing substantive and syntactic understanding. In their view, to develop substantive knowledge students also need to have an understanding of how that knowledge was developed. They further content that students cannot carry out an open-ended investigation if they do not have an understanding of the concept of evidence. Millar (2004) states that 'there are very few examples of successful implementation of extended investigation as part of the science curriculum in the context of 'mass education' where large numbers of teachers and students are involved' (p. 16). He argues that teachers find it difficult to come up with enough projects for students and so the investigations become routine and no longer what the curriculum intended. In some cases, the 'assessed investigation becomes almost the only investigation done' (p. 16).

While science investigation has the potential for students to learn about the nature of science, research shows that they struggle to do so. School science investigation appears most useful for syntactic learning when it is explicitly fore-fronted for learners and they actively reflect on and discuss the nature of science (Anderson, 2012; Hodson, 2009; Lederman, 2007). Akerson, Weiland, Pongsanon, and Nargund (2010) suggest that contextualising students' experiences of the nature of science in actual science investigation is as important as emphasising the nature of science during the investigation. Millar (2010a, 2010b) wisely suggests that teachers should have few and clear learning intentions for any one investigation, they should share them with the students, and at the end of the investigation ensure there is a reflective phase in which both students and teacher evaluate whether the intended outcomes have been achieved.

Research also suggests that expecting the school curricula and teachers to achieve both substantive and syntactic aims simultaneously often results in neither of these being achieved successfully (Osborne, 2014; Osborne & Dillon, 2008). Curricula are aspirational, written by expert science researchers and educators, and informed by science education research. Schools and teachers are given the task of implementation, which is dependent on teachers having an understanding as to why something is included or left out of the curriculum. Such understanding is often dependent upon the professional development they engage in when the curriculum is introduced. The actual implemented curriculum often depends on the purposes teachers personally see as important for science education, and their own beliefs about what ought to be learnt (Anderson, 2015). Other influential factors are the school context, access to resources, and the importance placed on national assessments to gauge the success of schools (Moeed, 2010). Millar (2011) suggests that science education requires a clear understanding by the teacher of the purpose for the learning experience and why it might be considered valuable at both individual and social levels.

1.7   Summary

In this chapter, we began by clarifying what we mean by school science investigation and considered the ways in which it differs from science investigation carried out by scientists. We highlighted the complexity of learning involved when students undertake science investigation, particularly open investigations. To clarify why science investigations might be worthy of inclusion in school science programmes, we considered the goals of science education. Many opportunities can be provided to learn about science ideas, develop investigative skills, and develop an understanding of the nature of science. In particular, we highlighted the multiple learning opportunities present in science investigation, including those that address current goals for science education learning: substantive knowledge, science processes, and the nature of science. Research into learning from science investigation suggests there are problems with meeting these multiple expectations. The intention to fulfil multiple purposes in any or every investigation often has the unintended consequence of none of the learning outcomes being achieved. Teacher and student beliefs about science teaching and learning and other contextual factors also appear to influence the learning that happens.

Delving into the literature leaves us questions about the practice of science investigation in our own New Zealand context. In particular, how do teachers perceive the nature and role of science investigation and how is it enacted in their classrooms? How do their students view science investigation and what are they learning through participation? What happens when teachers change their practice of teaching science investigation in the light of research? In the next chapter, we describe the New Zealand context and present our research design and methodology.

> **Questions to Consider**
> Osborne and Dillon (2008) recommend 'offering young people the best that is worth knowing' (p. 15). In light of the above discussion, what would you want to see included in a school curriculum for science? How do you see science investigation as contributing to that goal?

References

Abd-El-Khalick, F. (2012). Examining the sources for our understandings about science: Enduring conflations and critical issues in research on nature of science in science education. *International Journal of Science Education, 34*(3), 353–374.

Abd-El-Khalick, F., & Lederman, N. G. (2000). Improving science teachers' conceptions of the nature of science: A critical review of the literature. *International Journal of Science Education, 22*(7), 665–701.

Abrahams, I., & Millar, R. (2008). Does practical work really work? A study of the effectiveness of practical work as a teaching and learning method in school science. *International Journal of Science Education.* https://doi.org/10.1080/09500690701749305.

Akerson, V. L., Weiland, I., Pongsanon, K., & Nargund, V. (2010). Evidence-based strategies for teaching nature of science to young children. *Ahi Evran Üniversitesi Kırşehir Eğitim Fakültesi Dergisi, 11*(4), 61–78.

Allchin, D. (2011). Evaluating knowledge of the nature of (whole) science. *Science Education, 95,* 918–942.

Allchin, D. (2013). *Teaching the nature of science: Perspectives & resources.* St. Paul, MN: SHiPS Education Press.

Allchin, D. (2014). From science studies to scientific literacy: A view from the classroom. *Science & Education, 23*(9), 1911–1932.

Allchin, D. (2017). Beyond the consensus view: Whole science. *Canadian Journal of Science, Mathematics and Technology Education, 17*(1), 18–26.

Anderson, D. (2015). The nature and influence of teacher beliefs and knowledge on the science teaching practice of three generalist New Zealand primary teachers. *Research in Science Education, 45*(3), 395–423.

Anderson, D. (2012). Teacher knowledges, classroom realities: Implementing sociocultural science in New Zealand Year 7 and 8 classrooms. Unpublished doctoral dissertation, Victoria University of Wellington, New Zealand.

Anderson, D., & Moeed, A. (2017). Working alongside scientists. *Science & Education, 3*(4), 271–298.

Bolstad, R., & Hipkins, R. (2008). *Seeing yourself in science*. Wellington, NZ: New Zealand Council for Educational Research.

Bull, A., Gilbert, J., Barwick, H., Hipkins, R., & Baker, R. (2010). *Inspired by science*. Wellington: New Zealand Council for Educational Research.

Christodoulou, A., & Osborne, J. (2014). The science classroom as a site of epistemic talk: A case study of a teacher's attempts to teach science based on argument. *Journal of Research in Science Teaching, 51*(10), 1275–1300.

Driver, R., Leach, J., & Millar, R. (1996). *Young people's images of science*. McGraw-Hill Education (UK).

Fensham, P. J. (2004). Increasing the relevance of science and technology education for all students in the 21st century. *Science Education International, 15*, 7–26.

Ford, M. J., & Forman, E. A. (2006). Redefining disciplinary learning in classroom contexts. *Review of Research in Education, 30*(1), 1–32. https://doi.org/10.3102/0091732x030001001.

Goodrum, D., Rennie, L. J., & Hackling, M. W. (2001). *The status and quality of teaching and learning of science in Australian schools: A research report*. Canberra: Department of Education, Training and Youth Affairs.

Glaesser, J., Gott, R., Roberts, R., & Cooper, B. (2009). Underlying success in open-ended investigations in science: Using qualitative comparative analysis to identify necessary and sufficient conditions. *Research in Science & Technological Education, 27*(1), 5–30.

Hipkins, R., Bolstad, R., Baker, R., Jones, A., Barker, M., Bell, B. … Taylor, I. (2002). *Curriculum learning and effective pedagogy: A literature review in science education*. Wellington, NZ: Ministry of Education.

Hodson, D. (2009). *Teaching and learning about science: Language, theories, methods, history, traditions and values*. Boston: Sense.

Hodson, D. (2014). Learning science, learning about science, doing science: Different goals demand different learning methods. *International Journal of Science Education, 36*(15), 2534–2553. https://doi.org/10.1080/09500693.2014.899722.

Hodson, D., & Wong, S. L. (2014). From the horse's mouth: Why scientists' views are crucial to nature of science understanding. *International Journal of Science Education, 36*(16), 2639–2665.

Hodson, D., & Wong, S. L. (2017). Going beyond the consensus view: Broadening and enriching the scope of NOS-oriented curricula. *Canadian Journal of Science, Mathematics and Technology Education, 17*(1), 3–17.

Hume, A., & Coll, R. (2008). Student experiences of carrying out a practical science investigation under direction. *International Journal of Science Education, 30*(9), 1201–1228.

Lederman, N. G. (2004). Syntax of nature of science within inquiry and science instruction. In L. B. Flick & N. G. Lederman (Eds.), *Scientific inquiry and nature of science* (pp. 301–317). Dordrecht: Kluwer Academic.

Lederman, N. G. (2007). Nature of science: Past, present, and future. In S. K. Abell & N. G. Lederman (Eds.), *Handbook of research on science education* (pp. 831–879). Mahwah, NJ: Lawrence Erlbaum Associates.

Lederman, N. G., & Lederman, J. S. (2014). Is nature of science going, going, going, gone? *Journal of Science Teacher Education, 25*(3), 235–238. https://doi.org/10.1007/s10972-014-9386-z.

Millar, R. (2004). The role of practical work in the teaching and learning of science. In *High school science laboratories: Role and vision*. Retrieved from http://www.informalscience.org/images/research/Robin_Millar_Final_Paper.pdf.

Millar, R. (2010a). *Analysing practical science activities to assess and improve their effectiveness*. Hatfield: Association for Science Education. Retrieved from http://www.york.ac.uk/media/educationalstudies/documents/research/Analysing%20practical%20activities.pdf.

Millar, R. (2010b). Practical work. In J. Osborne & J. Dillon (Eds.), *Good practice in science teaching: What research has to say* (2nd ed., pp. 108–134). Maidenhead, UK: Open University Press.

Millar, R. (2011). Reviewing the national curriculum for science: Opportunities and challenges. *Curriculum Journal, 22*(2), 167–185.

Millar, R. (2015a). Experiments. In R. Gunstone (Ed.), *Encyclopaedia of science education* (pp. 418–419). Dordrecht: Springer.

Millar, R., & Osborne, J. (1998). *Beyond 2000. Science education for the future*. London: Nuffield Foundation.

Ministry of Education. (1995). *Investigating in science*. Wellington: Learning Media.

Moeed, A. (2010). *Science investigation in New Zealand secondary schools: Exploring the links between learning, motivation and internal assessment in year 11* (Unpublished doctoral thesis). Victoria University of Wellington, New Zealand.

Moeed, A. (2013). Science investigation that best supports student learning: Teachers' understanding of science investigation. *International Journal of Environmental and Science Education, 8*(4), 537–559.

Moeed, A. (2015). *Science investigation: Students' views about learning, motivation, and assessment*. Singapore: Springer. https://doi.org/10.1007/978-981-287-384-2.

Mortimer, E. F., & Scott, P. H. (2003). *Meaning making in secondary science classrooms*. Maidenhead: Open University Press.

National Research Council. (2012). *A framework for K-12 science education: Practices, crosscutting concepts, and core ideas*. Washington, DC: Committee on a Conceptual Framework for New K-12 Science Education Standards. Board on Science Education, Division of Behavioral and Social Sciences and Education.

Osborne, J. (2014). Teaching scientific practices: Meeting the challenge of change. *Journal of Science Teacher Education, 25*(2), 177–196.

Osborne, J. (2017). Going beyond the consensus view: A response. *Canadian Journal of Science, Mathematics and Technology Education, 17*(1), 53–57.

Osborne, J., Collins, S., Ratcliffe, M., Millar, R., & Duschl, R. (2003). What "ideas-about-science" should be taught in school science? A Delphi study of the expert community. *Journal of Research in Science Teaching, 40*(7), 692–720.

Osborne, J., & Dillon, J. (2008). *Science education in Europe: Critical reflections* (Vol. 13). London: The Nuffield Foundation.

Roberts, R. (2009). Can teaching about evidence encourage a creative approach in open-ended investigations. *School Science Review, 90*, 31–38.

Schwab, J. J. (1962). The teaching of science as enquiry. In J. Schwab & P. Brandwein (Eds.), *The teaching of science* (pp. 3–102). Cambridge, MA: Harvard University Press.

Schwab, J. J. (1964). The structure of the natural sciences. In G. W. Ford & L. Pugno (Eds.), *The structure of knowledge and the curriculum* (pp. 31–38). Chicago: Rand McNally.

Stuckey, M., Hofstein, A., Mamlok-Naaman, R., & Eilks, I. (2013). The meaning of 'relevance' in science education and its implications for the science curriculum. *Studies in Science Education, 49*(1), 1–34.

Toplis, R. (2012). Students' views about secondary school science lessons: The role of practical work. *Research in Science Education, 42*(3), 531–549.

Tytler, R. (2007). *Re-imagining science education: Engaging students in science for Australia's future*. Melbourne: Australian Council of Education Research.

Tytler, R., & Osborne, J. (2012). Student attitudes and aspirations towards science. In *Second international handbook of science education* (pp. 597–625). Netherlands: Springer.

Windschitl, M., Thompson, J., & Braaten, M. (2007). Beyond the scientific method: Model-based inquiry as a new paradigm of preference for school science. *Science Education, 92*(5), 941–967.

Wong, S. L., Hodson, D., Kwan, J., & Yung, B. H. W. (2008). Turning crisis into opportunity: Enhancing student-teachers' understanding of nature of science and scientific inquiry through a case study of the scientific research in severe acute respiratory syndrome. *International Journal of Science Education, 30*(11), 1417–1439.

Chapter 2
The New Zealand Context and Research Design

2.1 Science in the New Zealand Curriculum

In this chapter, we describe the research project that forms the basis of this book. We begin by outlining the New Zealand educational context. We then present an overview of the New Zealand curriculum before moving on to describe the research design and methodology.

Schooling in New Zealand is compulsory from ages six to sixteen, although most children start school on their fifth birthday. Primary (elementary) schooling, therefore, usually begins at age five in Year 1 and continues through to Year 8 in most English medium schools, although Māori or Pasifika medium education is also an option. Most Year 7 and 8 students attend either a primary school for Years 1–8 or an intermediate school for Years 7 and 8 after attending a primary school that just caters for Years 1–6. Secondary (high school) schooling is from Years 9 to 13 but there are some Year 7–13 schools in rural areas.

Science is one of the eight learning areas encompassed in the New Zealand curriculum and is compulsory from Years 1 to 10 (ages 5–15). Science courses in the lower secondary school are usually general science courses. In senior secondary school, most schools offer the traditional physics, chemistry and biology disciplines though a few offer a general science programme.

Teaching and learning science in New Zealand primary schools is an ongoing concern. Students' experience of science investigation is limited as most teaching is information based (Education Review Office, 2012). International studies show that New Zealand primary teachers are more likely to be generalists, are less confident to teach science, and are less likely than many of their international counterparts to have undertaken recent professional development in science. New Zealand primary school students spend less time studying science than in many other countries, and there is less focus on experiments and investigations; students more often look up ideas and science information (Caygill, Singh, & Hanlar, 2016). Two likely reasons are cited for this: one is that the New Zealand curriculum is non-prescriptive and the

© Springer Nature Singapore Pte Ltd. 2018
A. Moeed and D. Anderson, *Learning through School Science Investigation*,
https://doi.org/10.1007/978-981-13-1616-6_2

nature of implemented curriculum varies at both the school and class level; second, the introduction of National Standards of assessment for literacy and numeracy in primary schools has had the effect of reducing curriculum time spent on science in schools. Science is the most commonly integrated in primary schools with other curriculum areas in a generic inquiry approach. Students use an inquiry cycle, usually adapted for their school and most commonly aligned to information literacy approaches, to find the answers to their own questions (Boyd & Hipkins, 2012). A review of primary science found that in many schools, science learning was lost in an integrated approach (Hipkins et al., 2002).

Secondary schools in New Zealand have undergone two systemic changes in recent times, first, the introduction of the National Certificate of Educational Achievement (NCEA) in 2002, and second, the introduction and implementation of a new curriculum in 2007. Both these changes have influenced teaching, learning and assessment of science investigation. The curriculum prioritises teaching about the nature of science and investigating in science. It aims for students to experience different approaches to the investigation including classifying and identifying, pattern seeking, exploring, investigating models, fair testing and making things or developing systems. However, assessment of science investigation as *fair testing* has constrained both teaching and learning investigation in Year 11 (Hume & Coll, 2008; Moeed & Hall, 2011). This is not unexpected as similar issues arose when internal assessment of investigation was introduced in the UK (Roberts & Gott, 2004). In 2011, the NCEA achievement standards, which are used for high-stakes assessment in Years 11–13, were aligned with the curriculum by the Ministry of Education and it appears that the achievement standards have become the default curriculum in senior secondary school (Tewkesbury, 2017).

The *New Zealand Curriculum* sets out the values and key competencies that students should develop as part of their education if they are to become lifelong learners. It also describes broad achievement objectives for eight learning areas. The multiple purposes of the science learning area include: preparing students for a career in science, developing practical knowledge of how things work, developing science literacy to enable informed participation in a science discussion, and developing skills in scientific thinking and knowledge (Bull, Gilbert, Barwick, Hipkins, & Baker, 2010). The current science curriculum in New Zealand states:

> In **science**, students explore how both the natural physical world and science itself work so that they can participate as critical, informed, and responsible citizens in a society in which science plays a significant role. (Ministry of Education, 2007, p. 17)

The curriculum expectation is that by studying science, students:

- develop an understanding of the world, built on current scientific theories
- learn that science involves particular processes and ways of developing and organising knowledge and that these continue to evolve
- use their current scientific knowledge and skills for problem solving and developing further knowledge

- use scientific knowledge and skills to make informed decisions about the communication, application, and implications of science as these relate to their own lives and cultures and to the sustainability of the environment. (p. 28)

The New Zealand Curriculum prioritises learning about how science works by making the Nature of Science strand the overarching strand. The other four strands—Living World, Material World, Physical World, and Planet Earth and Beyond—provide the context for learning about the nature of science. The Nature of Science strand is given primacy in order to meet the goal of developing scientifically literate citizens. It is divided into four sub-strands:

Understanding about science

Learn about science as a knowledge system: the features of scientific knowledge and the processes by which it is developed; and learn about the ways in which the work of scientists interacts with society.

Investigating in science

Carry out science investigations using a variety of approaches: classifying and identifying, pattern seeking, exploring, investigating models, fair testing, making things, or developing systems.

Communicating in science

Develop knowledge of the vocabulary, numeric and symbol systems, and conventions of science and use this knowledge to communicate about their own and others' ideas.

Participating and contributing

Bring a scientific perspective to decisions and actions as appropriate.[1]

Teachers are expected to develop programmes of learning that address: the requirements of the science content area; the Nature of Science strand and its overarching place in the curriculum; the key competencies that students are expected to develop as lifelong learners; and the alignment of the supporting science resources. To make this complex task more achievable, five science capabilities for citizenship have been proposed, and are outlined below. The purpose of these five capabilities is for teachers to use them to help students develop a functional understanding about the nature of science. These science capabilities have been developed in New Zealand (Bull, 2015; Hipkins & Bull, 2015) and are similar to the requirements of the Framework for K-12 Standards (National Research Council, 2012) in the United States. Osborne (2014) theorises that the US standards provide a strong representation of the current understanding of the nature of science as a social and cultural practice. His view is that adopting a framework of practices will improve communication about the meaning of the practices of science amongst professional science educators. The long-term vision is to improve practice in the classroom. The science capabilities for citizenship reflect these views and can be accessed through the Ministry of Education website (http://scienceonline.tki.org.nz/Science-capabilities-for-citizenship). Table 2.1 sets out the capabilities and the

---

[1]http://nzcurriculum.tki.org.nz/The-New-Zealand-Curriculum/Science/Achievement-objectives#collapsible1.

Table 2.1 The five science capabilities for citizenship

Capability	What learners do	What learners are intended to learn
Gather and interpret data	Learners make careful observations and differentiate between observation and inference	*Science knowledge is based on data derived from direct, or indirect, observations of the natural and physical world and often includes measuring something. An inference is a conclusion you draw from observations—the meaning you make from observations. Understanding the difference is an important step towards being scientifically literate*
Use evidence	Learners support their ideas with evidence and look for evidence supporting others' explanations	*Science is a way of explaining the world. Science is empirical and measurable. This means that in science, explanations need to be supported by **evidence** that is based on, or derived from, observations of the **natural world***
Critique evidence	Not all questions can be answered by science	*In order to evaluate the trustworthiness of data, students need to know quite a lot about the qualities of scientific tests*
Interpret representations	Scientists represent their ideas in a variety of ways, including models, graphs, charts, diagrams and written texts	*Learners think about how data are presented and ask questions such as:* • *What does this representation tell us?* • *What is left out?* • *How does this representation get the message across?* • *Why is it presented in this particular way?*
Engage with science	This capability requires students to use the other capabilities to engage with science in 'real life' contexts	*This capability involves students taking an interest in science issues, participating in discussions about science and at times taking action*

Note The information in this table is taken from Five Capabilities for Citizenship: (http://scienceonline.tki.org.nz/Science-capabilities-for-citizenship)

website provides a number of suggested resources for teachers that can be used to support their development.

These science capabilities are relevant to the research reported in this book because they are presented as a way of addressing the Nature of Science strand that meets the intended purpose of the curriculum, i.e. to be ready, willing and able to engage with science as part of a society in which science holds increasing importance.

2.2 Research Design and Methodology

As the previous chapter has argued, while we know that teachers often have multiple goals for learning from science investigation, little is known about how their perceptions of science investigation change across the primary and secondary education sectors, how perceptions of nature and purpose of investigation vary, how and why teachers use them in their classes, and what students learn from them. There is some evidence to suggest that students perceive science investigation to be an enjoyable part of science, but don't think they learn from it, or if they do, that the learning is the direct result of their observations—'what happens when …'

We developed the following research questions to guide our research:

1. How do participating teachers conceptualise science investigation, and its educational role, within school science?

   a. What purposes do they see for school science investigation?
   b. What approaches to science investigation do they practice?
   c. How do the types of school science investigation that students experience vary in different educational contexts, primary and secondary?

2. What types of school science investigation do students experience in the different educational contexts, primary and secondary?
3. Which characteristics of science investigation, or ways of presenting science investigation, facilitate the development of students' substantive and nature of science understandings?
4. In what ways does research-informed change in practice influence student learning through science investigation?

Research Design

We chose to use a qualitative interpretive paradigm for our research. Quantitative methods, like questionnaires and surveys, have been useful to examine views about practical work (e.g. Beatty & Woolnough, 1982; Thompson, 1975). They are comparatively cheap and could give indications of patterns and trends across sectors. However, our interest in teachers' and students' understanding of science investigation and their experience of it lies in the interpretive or qualitative paradigm, where knowledge is considered subjective and socially embedded (Cohen, Manion, & Morrison, 2011). Interpretive research aims to communicate to its audience the understandings developed by observing and recording the everyday life of the participants (Merriam, 2001; Patton, 2002). As Grbich (2007) suggests, 'Multiple realities are presumed, with different people experiencing these differently' (p. 8). We wanted to gather a first-hand, in-depth and detailed picture of the variety of students' and teachers' perceptions and experiences of science investigation in New Zealand schools as reflected in our choice of methodology and data gathering tools.

We selected a multiple case study approach. Classroom-based case studies that provide a detailed and in-depth understanding of the multiple variables of the classroom environment and describe the details of student–teacher interactions and learning are few (Baker & Jones, 2005). They can do much to enhance our understandings of science learning in classrooms, revealing 'how all the parts work together to form a whole' (Merriam, 2001, p. 6). Cases are bounded systems (Merriam, 2001). The cases in this study were bounded by level of schooling (the stage of schooling during which the investigations occurred), duration (all investigations occurred within one unit or topic of work in science), place (the environments, normally the school laboratory or classroom in which the teaching occurred) and the participants, the teacher and their class. The cases were also bounded by the learning area; only science lessons were observed.

Multiple case studies are beneficial in that the variety allows a more compelling interpretation (Merriam, 2001). Stake (2006) argues that a multiple case study can elucidate the ways that a phenomenon can operate in different situations. He proposes that while cases could be selected statistically to represent the population of cases, education and teaching, in particular, has a different purpose in studying a variety of cases. The different cases each add to the understanding of the phenomenon. It is helpful in education to build an understanding of the interrelated factors in different contexts that influence the way a phenomenon, in this study science investigation, can be enacted and perceived. A multiple case study approach was therefore chosen as it best met the purposes of the study.

Data collection occurred in two phases. The data that were collected about current practice could then be used to inform teachers' decisions about aspects of practice that they wished to improve in. We, therefore, followed the teachers for two years. The research design is shown in Fig. 2.1.

Our goal was for the teachers to have ownership of any change in practice. We wanted them to have opportunity and time to reflect on the findings and identify elements that they themselves wanted to work on and improve. Data collection for Phase One occurred in the first year of the study. These data were analysed and general findings presented to the teachers before their summer break at the end of the first year. At this time, they were also provided with a research article which they could use to assess and improve the effectiveness of their use of school science investigation (Millar, 2010). The summer break allowed the teachers time for this reflection and professional reading. After the break, at the start of the new teaching year, the teachers all met together again. They discussed their beliefs about science investigation, as findings from interviews in Phase One had revealed very broad interpretations of science investigations which we wanted to probe further. By this time, teachers had identified aspects of their practice that would form the focus for change and development in Phase Two. Researchers then gathered further research articles for each teacher relevant to the changes they wanted to make. Data collection for Phase Two occurred in the second year of the study.

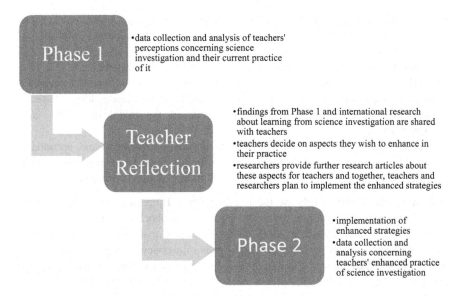

Fig. 2.1 The research design

Data Collection Methods and Analysis

A range of data sources informed the study in both phases and are presented in the next section. Slight variations between cases are explained in the reports of each case study in Chaps. 3–5. For consistency, one researcher gathered all data for the primary teachers and another for the secondary teachers.

The following data sources were used during the two years of this research project:

- A semi-structured interview with each teacher based on an example of a science investigation that they had successfully implemented (see Appendix 2.1).
- The unit plan pertaining to the topic taught during data collection.
- A group discussion with participating teachers about the nature and purpose of school science investigation.
- Classroom observations of three lessons for each class where teachers had implemented an investigative approach. The lessons were audio recorded, an observation schedule completed (see Appendix 2.2), and running records/field notes taken to gain insight into learning opportunities.
- Following each observed lesson, a focus group interview conducted with five/ six students selected by negotiation with teachers to provide a range of ethnicities/abilities/interest in science. Interviews inquired into students' perceptions about their substantive and syntactic learning and their perceptions of the purpose and role of school science investigation. For the new entrant class, individual interviews with students, rather than a focus group, were recorded.

- Artefacts resulting from the research investigation, including student work, short video-clips and audiorecordings of conversation with students, collected as appropriate during the investigation.
- Teacher and researcher reflections at the end of the data collection.
- At the end of Phase One, interviews with primary teacher and secondary teacher focus groups using printed comments designed to prompt discussion on aspects that required clarification following analysis of the initial interviews (see Appendix 2.3).
- Questionnaire completed by students in the middle primary and secondary school classes at the end of the unit which included questions relevant to their topic, and also a set of common questions about science investigations.

Data were coded inductively by the research team. Emerging codes and themes were discussed and examples and exemptions cross-checked using a process of constant comparison (Merriam, 2001). The Science Capabilities for Citizenship[2] were used deductively to analyse focus group interviews, questionnaires, and observations. In each case, accuracy and common understanding were required for consistency in allocating the codes. A constant comparison process (Merriam, 1998) was used in open coding of phrases and words used by participants. Millar's (2010) framework was used to analyse student surveys and these are reported in Chap. 5.

Participants and Cases

Cases were selected purposively. In the case of the primary schools, the two teachers invited to participate were known to teach science regularly and worked at different levels in the primary sector. Both teachers had participated in 2010 in the Primary Science Teacher Fellowship scheme[3] administered by the Royal Society of New Zealand, designed to enhance teaching of science within the school community, and so had an interest in teaching science. The schools were also chosen as the researchers had an established relationship with the teachers or schools and for their location, allowing the researchers to visit frequently for interviews and observations. It was originally intended to have two participating teachers from each sector, but when the Head of Science in the secondary school approached teachers in her Year 9 and 10 classes, three were keen to be involved and so it was agreed they should all participate. Year 10 (student age 14) is the stage to which science is compulsory in NZ schools and, therefore, was set as the upper limit for the second year of the study. The primary school case study was carried out with two teachers in separate schools. The principals, teachers and their students

_____

[2](http://scienceonline.tki.org.nz/Science-capabilities-for-citizenship/Introducing-five-science-capabilities).

[3](www.royalsociety.org.nz/teachinglearning/science-teaching-leadership-programme).

provided informed consent as required by the ethics committee. For the primary students, informed consent from both students and caregivers was obtained. A letter and photo of the researcher explaining the project and consent form in child-friendly language were sent home with the information sheet and consent form for parents or caregivers.

Details of the setting and participants are provided in the case studies in Chaps. 3 and 4. In all five cases, the teachers and their classes formed the participants, two primary classes being from different schools (New entrants, aged 5; and Year 5/6 students, aged 9/10 for both phases) and three secondary classes from the same school (Year 9, students aged 13 for Phase One and Year 10, students aged 14 for Phase Two).

2.3 Summary

We began this chapter by presenting the New Zealand context of our research and the mandated requirements for science in the New Zealand curriculum. We described the current concerns about teaching and learning science in primary and secondary schools. Focussing on the place of science investigation within the curriculum, we provided the details of the Nature of Science strand of the curriculum and the aims for the 'investigating in science' sub-strand. Next, we explained the recently proposed Science Capabilities for Citizenship, a framework we have used in the present study. The second part of this chapter provided the case study research design and the research questions. The research questions provided the justification for having two phases, as well as the participant selection, data sources and analytic approaches. The chapter concluded with a description of cases and participants.

In Chap. 3, we present case studies of the two participating primary teachers' existing practices that support student learning through science investigation.

Questions to Consider

1. What is your current understanding of science capabilities? In what ways do you think the science capabilities support students' learning about the nature of science investigation?
2. Critically evaluate the study design. What strengths and weaknesses do you see in the research design and methodology? If you were designing this research, what changes would you make and why?

Appendix 2.1: First Interview with Teachers

Teachers are asked to bring a science investigation that they have done with students

1. Thanks so much for making this time to chat today. We are very much just starting out on this pilot stage of the project. We are so grateful that you are prepared to be involved. What would you like to get out of this project?
2. Do you mind telling me a bit about your teaching experience?
   Possible probes:

   - How long have you been teaching?
   - What levels have you taught and currently teach?
   - What made you become interested in teaching science?

3. Please tell me about the science investigation that you have brought along.

   (a) Why did you bring this one?
   (b) What was your experience with using it?
   (c) Did you think it was a good investigation? Why?

   Possible probes:

   - What did you want the students to learn from doing it?
   - What was the students' role in this investigation? (Give time before probing)

     – to decide the question;
     – to choose the equipment to use;
     – to decide what to observe or measure;
     – to decide how to summarise and analyse the data, etc.

4. (if not brought up in 3) What do you see as the key elements of any science investigation?
5. Some researchers say that school science is nothing like a real science investigation done by scientists—what do you think?

   - Do you agree?
   - Do you think school science investigations should be like real science investigations?

6. Do you think all students/children should do science investigations? Why? Why not?
7. What guides your thinking when choosing and planning a science investigation?
8. How do you see science investigation in relation to science education?
9. If you could teach any way you wanted to, how would you like to teach science?

   (a) What would you most like to teach about?
   (b) How would you like to teach it?

Appendix 2.2: Sector Interview Provocations

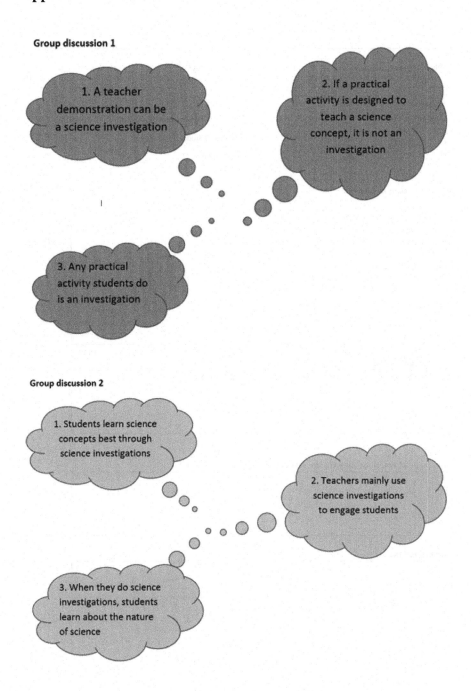

Appendix 2.3: Observation Schedule

Observation Schedule:	Date:	Day:	Time

Classroom environment:

Physical aspects:

- Layout
- Lighting
- Room Temperature

Teacher welcomes students to class

Number of students in class

Demeanour

Lesson overview

Start of the lesson: topic/focus, style
Uses an advance organiser:
Learning intentions shared: Written/Verbal
Makes links to the previous lesson:
Checks homework:

Contextualisation

Looking for connecting with students; students' world, culture

Evidence of relationship with the class

Investigation:

Type: _____

Skill involved and developed:

Taught   Practised   Applied

How is the investigation introduced?

Teacher led   Demonstration   Posed as a problem   Student driven

Instructions

Written/ verbal/on hand out/From the text book

Student involvement

How are students involved?
Individually/in groups [teacher or self-selected]
Decision about what to:

- Investigate
- Choose equipment
- Decide what to measure
- Decide how to record results
- Decide how to analyse and summarise
- Were they given time to think?
- Critique their own design
- Critique others design

How was the investigation concluded?
Reflection yes/No
What was the focus for the reflection?

Quality of Engagement:
(Checked three times in the lesson)

Enthusiastic: (Keenness/eager to start)	Most	some	Few
Perseverance: (Carrying it through)	Most	some	Few
Attentive: (Behaving within the guidelines of class expectation)	Most	some	Few
On task behaviour:	Most	some	Few

Researcher's view about what was learnt?

What was the focus: Conceptual/Procedural/Skill development

What do you think students learnt?

Evidence:

References

Beatty, J. W., & Woolnough, B. E. (1982). Practical work in 11-13 science: The context, type and aims of current practice. *British Educational Research Journal, 8*(1), 23–30.

Baker, R., & Jones, A. (2005). How can international studies such as the international mathematics and science study and the programme for international student assessment be used to inform practice, policy and future research in science education in New Zealand? *International Journal of Science Education, 27*(2), 145–157. https://doi.org/10.1080/0950069042000276695.

Boyd, S., & Hipkins, R. (2012). Student inquiry and curriculum integration: Shared origins and points of difference (Part A). *Set: Research Information for Teachers,* (3), 15.

Bull, A. (2015). Capabilities for Living and Lifelong Learning: What's Science Got to Do with It? New Zealand Council for Educational Research. PO Box 3237, Wellington 6140 New Zealand.

Bull, A., Gilbert, J., Barwick, H., Hipkins, R., & Baker, R. (2010). *Inspired by science.* Wellington: New Zealand Council for Educational Research.

Caygill, R., Singh, S., & Hanlar, V. (2016). *TIMSS 2014-15 Science Year 5. Trends over twenty years in TIMSS. Findings from TIMSS2014-15.* Retrieved 1 October, 2017 from http://www.educationcounts.govt.nz/__data/assets/pdf_file/0003/180390/TIMSS-2014-15-Science-Year-5-Trends-over-20-years-in-TIMSS.pdf.

Cohen, L., Manion, L., & Morrison, K. (2011). *Research methods in education* (7th ed.). Abingdon, Oxon: Routledge.

Education Review Office. (2012). Science in the New Zealand Curriculum Years 5 to 8. http://www.ero.govt.nz/National-Reports/Science-in-The-New-Zealand-Curriculum-Years-5-to8-May-2012/. Accessed 13 Dec 2013.

Grbich, C. (2007). *Qualitative data analysis: An introduction.* London: SAGE.

Hipkins, R., Bolstad, R., Baker, R., Jones, A., Barker, M., Bell, B., et al. (2002). *Curriculum, learning and effective pedagogy: A literature review in science education.* Wellington, NZ: Ministry of Education.

Hipkins, R., & Bull, A. (2015). Science capabilities for a functional understanding of the nature of science. *Curriculum Matters, 11,* 117–133.

Hume, A., & Coll, R. (2008). Student experiences of carrying out a practical science investigation under direction. *International Journal of Science Education, 30*(9), 1201–1228.

Merriam, S. B. (1998). *Qualitative research and case study applications in education.* Revised and expanded from "Case Study Research in Education." Jossey-Bass Publishers, 350 Sansome St, San Francisco, CA 94104.

Merriam, S. B. (2001). *Qualitative research and case study applications in education.* San Francisco: Jossey-Bass.

Moeed, A., & Hall, C. (2011). Teaching, learning and assessment of science investigation in year 11: Teachers' response to NCEA. *New Zealand Science Review, 68*(3), 95–102.

Millar, R. (2010). *Analysing Practical Science Activities to assess and improve their effectiveness.* Hatfield: Association for Science Education. Retrieved from http://www.york.ac.uk/media/educationalstudies/documents/research/Analysing%20practical%20activities.pdf.

Ministry of Education. (2007). *The New Zealand curriculum.* Wellington: Learning Media.

National Research Council. (2012). A framework for K-12 science education: Practices, crosscutting concepts, and core ideas. Washington, DC.: Committee on a Conceptual Framework for New K-12. Science Education Standards. Board on Science Education, Division of Behavioral and Social Sciences and Education.

Osborne, J. (2014). Teaching scientific practices: Meeting the challenge of change. *Journal of Science Teacher Education, 25*(2), 177–196.

Patton, M. Q. (2002). *Qualitative evaluation methods* (2nd ed.). Thousand Oaks, CA: Sage.

Roberts, R., & Gott, R. (2004). Assessment of Sc1: Alternatives to coursework? *The School Science Review, 85*(131), 103–108.

Stake, R. E. (2006). *Multiple case study analysis.* New York: The Guildford Press.

Tewkesbury, M. (2017). *An analysis of National Certificate of Educational Achievement chemistry courses in terms of curriculum, pedagogy and assessment: With comparison to the International Baccalaureate Diploma.* A thesis submitted to Victoria University of Wellington in fulfilment of the requirements for the degree of Doctor of Philosophy, September, 2017.

Thompson, J. J. (Ed.). (1975). *Practical work in sixth form science.* Oxford, UK: Department of Educational Studies, University of Oxford.

Chapter 3
Science Investigation in Primary School

3.1 The NZ Primary Context and Curriculum

As described in Chap. 2, *The New Zealand Curriculum* (Ministry of Education, 2007) positions the Nature of Science strand as the 'overarching unifying strand' (p. 28) that requires learning for all students to Year 10. The Nature of Science strand provides objectives guiding student learning about scientific investigation and their appreciation of epistemological and sociological aspects of science as a discipline. There is a strong focus on play, exploration and talk in the achievement objectives that describe the learning that is intended for the first four years of science education. For example, the Investigating in Science sub-strand for Level 1/ 2, in setting out expectations for the first four years of schooling, states that students should 'extend their experiences and personal explanations of the world through exploration, play, asking questions and discussing simple models' (p. 47). As students move through the primary school system there is a move towards more systematic investigation, as well as developmentally progressing from a personal focus on self to more collaborative approaches. By Level 3/4, students should 'build on prior experiences, working together to share and examine their own and others' knowledge' and 'Ask questions, find evidence, explore simple models and carry-out appropriate investigations to develop simple explanations' (p. 48). At the same time, the contextual strands should provide 'contexts for learning' (p. 29). Young primary school students are expected to 'seek and describe simple patterns in physical phenomena' and 'observe, describe, and compare physical and chemical properties of common materials and changes that occur when materials are mixed, heated, or cooled' (p. 47).

While the curriculum sets out these broad expectations for science education, the complex task of weaving together a programme that supports students to develop both types of science learning outcomes (substantive and syntactic) is left to individual schools. Primary schools, as a whole, are expected to have sufficient understanding to make 'decisions about how to give effect to the national

© Springer Nature Singapore Pte Ltd. 2018
A. Moeed and D. Anderson, *Learning through School Science Investigation*,
https://doi.org/10.1007/978-981-13-1616-6_3

curriculum in ways that best address the particular needs, interests, and circumstances of the school's students and community' (Ministry of Education, 2007, p. 37). Ultimately, it falls to the classroom teacher to design and implement opportunities that support student learning in science. Yet, New Zealand primary school teachers, like many of their colleagues internationally, are generalists, and few have a background in science (Bull, Gilbert, Barwick, Hipkins, & Baker, 2010). A lack of confidence, most often attributed to a perceived lack of content knowledge, has been identified as impacting on effective science teaching in successive reviews of primary school science education in New Zealand (Education Review Office, 2004, 2010, 2012). Primary teachers often see science as requiring complex and expert knowledge and place authority on books and the Internet (Appleton, 2006). What teachers believe about what science is impacts on the way they teach it (Anderson, 2015). Reviews of New Zealand primary school teaching practice show that knowledge-based programmes relying on written information are more common than those where students participate in practical investigation, discussion and thinking (Education Review Office, 2012). In international comparisons, New Zealand Year 5 teachers conduct fewer experiments and investigations in science than in other OECD countries. They also receive less professional development in science (Chamberlain & Caygill, 2012).

In this chapter, we examine the beliefs about science investigation held by the two primary teachers who participated in the study and consider how these beliefs translated into their practice. A description of each teacher's practice of science investigation during Phase 1 of the project is provided along with findings from analysis of observations. We identify strategies these teachers were already incorporating into their teaching in the three key areas that were the focus of the study: supporting students to investigate scientifically; connecting students' practical experiences to key scientific concepts and making learning about the nature of science explicit. Finally, we consider how practices impacted student learning through an analysis of student interviews and work samples, discussing some of the challenges that were identified during each set of observations.

3.2 Participants and Contexts

The two teachers who participated in this study are not perhaps typical of New Zealand primary teachers in that they had a special interest in science and, as a result, participated in a long-term professional development project, the Primary Science Teacher Fellowship.[1] This programme aimed to build teachers' confidence

---

[1]This programme was part of the New Zealand Science, Mathematics and Technology Teacher Fellowship Scheme funded by the Ministry of Business, Innovation and Employment and administered by the Royal Society of New Zealand. It has been adapted to include junior high school science teachers and is now known as the Science Teacher Leadership Programme (http://www.royalsociety.org.nz/teaching-learning/science-teaching-leadership-programme/).

and leadership in science. The teachers had both worked full time for six months alongside scientists in science-based organisations and during that time had participated in several professional development days facilitated by science teacher educators that were designed to help participants connect their experiences to their classroom practice. The two teachers were purposively selected for the study as they were likely to be teaching science regularly and had an understanding of scientific investigation and the aims and expectations of *The New Zealand Curriculum* (Ministry of Education, 2007).

Alison had been teaching for 15 years full time with a 10-year break when her children were young. She also had experience in early childhood education. Although she had some experience of teaching at years 7 and 8, most of her teaching experience has been in the junior primary area, with five years at the new entrant level (age 5 years). Her experience in the Primary Science Teacher Fellowship involved working alongside geologists. Alison regularly engages her students in practical science experiences and has worked to raise awareness of science in the school as a whole, creating school-wide opportunities where science could be fostered. For example, she organised the building of a butterfly enclosure at the school. Students from Alison's class were involved in the design and choice of plants for the enclosure. They regularly observe the butterflies and are involved in a national monitoring project, tagging monarch butterflies and releasing them. Her class also completed a study of the monarch butterfly life cycle. Student work, including work in science, is celebrated through Alison's class blog and through vibrant classroom displays sharing students' ideas, individual writing, drawing, teacher-written class stories, and photos of their activities. Alison uses science experiences to provide joint opportunities for learning in numeracy and literacy.

Students in Alison's class were mostly New Zealand European from high socio-economic homes. Parents had high expectations of students, and the school was supportive of the strong science focus in Alison's class. This was evident by enthusiastic parent comments made to the researcher during observations, the number of students who brought science-related objects from home during both units, and the fact that all parents responded with consent for their child's participation in the study. In the first year of the study, there were 16 students in the class, all aged five years—nine boys and seven girls. Several were very new to school but most students had been at school for one or two terms. Observations were carried out towards the end of the third term. In the second year, the class comprised 18 students—ten boys and eight girls. Again, the majority had been at school only for a term or so and there were several who were very new.

The second teacher, Patsy, had been teaching for 24 years. While she had some time out of teaching and had worked part-time for several years, at the time of the study she had been teaching eight years full time at the Year 5/6 level (age 9–10 years). She had previous experience teaching junior classes. Patsy had taken biology at secondary school, but her interest in teaching science was largely because of students' positive and curious response to science and the world around them. She felt science was 'kid-friendly': 'You can pick a seed head and they think it is really cool. They are really interested, and I think that has got even more so now because it seems that children have lost that connection with their environment.' She felt science was a way to reconnect them to the world around them. Her Primary Science Teacher Fellowship saw her spending six months working with environmental scientists.

Patsy's school for Phase 1 of the project was in a mid range socio-economic community with some students coming from quite low-income homes and some from middle-income homes. The school has a reputation in the local community for supporting students with special needs and behavioural difficulties. Patsy had one such student in her Year 5/6 class of 24 students. The class had roughly equal numbers of boys and girls. An adult helper was present for two of the three sessions observed. Patsy was responsible for science in her level of the school, making sure other teachers had the necessary materials and plans as well as preparing for and teaching science in her own classroom. The students seemed keen and pleased to be doing practical science. They were generally well-organised, self-managing and responsible in carrying out practical work.

3.3   Teachers' Beliefs About Science Investigation

Both teachers tended to use the term 'science investigation' loosely to refer to any practical work in science, but also in a specialised way that involved answering students' questions in practical scientific ways.

For Alison, activities that seemed to have known conceptual goals such as making prescribed models, teacher demonstrations, student gathering of data about how shadows change during the day, were all included among her examples of investigations. Practical experiences and observational opportunities were also discussed as investigations. On deeper probing, the use of student questions and ownership appeared to be a special component Alison associated with science investigation, despite her broad application of the term. Another key element was student interest in, and relevance of the topic: 'It has to be something that is relevant to the children … something interesting and something that they want to find out about.'

Similarly, Patsy referred to a range of practical science activities as investigations. An engaging topic and student development of questions were important features for her also: 'That whole idea of observing and asking questions, it is really important. Getting the kids to actually understand how to do that.' She felt the important learning from investigation was about scientific processes 'testing and having a clear idea of what you are trying to observe and look for.' Inference was also important: 'If I have seen this then what does that make me think about?' These views were connected to her experiences of science in her fellowship; she wanted her students to understand what made science unique as a discipline: 'what makes being a scientist different … when you are being a scientist what do you do?' She felt a science investigation created opportunities for collaboration that replicated those in the science world: 'they [students] work together on it – the whole collaborative thing – and there is a reason for that ... you help each other; it is a good way to work and that is why scientists often do work together.'

In an effort to tease out more of the range of beliefs that the teachers held about investigations, a concept cartoon activity (see Appendix 2.2) was presented to both teachers together and the ensuing discussion recorded and analysed. These conversations led to some agreement that science investigations are open-ended. Students need to be able to explore a situation or phenomenon rather than be shown how it works. They agreed that school science investigations should engage students in doing science and that student questions are central. Shulman (1986) suggests that teachers, whether intentionally or not, by their actions and what they include and value in disciplinary teaching, convey messages about the nature of the discipline. The teachers appeared to be aware that they bore this responsibility for science, and wanted their students to have an authentic experience of science through their investigations:

Patsy   They *do* something with it, I guess. If the kids just think, 'oh I just need to *know*', then that's not really being a scientist.
Alison (about teacher demonstrations): I think students would be short changed ... they'd be thinking, well, science, is that the teacher showing me things?
Patsy   If you're wanting them to realise that science is about investigating their world and understanding it better and how the world works, then you have to get your hands in the world ... You have to get in there and do it.

These teachers saw science investigations as purposeful activity; they were more than play or a means of engaging students. As Patsy commented:

You can imagine mixing oil and water and stuff – the kid who does it and really looks to see what happens, and the kid that just wants to stir it all up and keep slopping things and playing ... they need to think about what they're trying to find out to make it an investigation.

They expressed tensions between teaching about the nature of science and teaching for an understanding of science concepts through science investigation:

We can say science is about trying to explain things and we infer explanations from what we see and might be related to, and [theories] could be challenged in the future but I think most people, children and adults sort of think, what's the right answer? What's really happening?... And so you're kind of on that tightrope really with the tension because sometimes you do want to say, well we're pretty sure that's how it is. (Patsy)

These views reflect the multiple purposes that exist for including investigations in a science programme and the dilemmas teachers face in trying to address them. Millar (2011) and Hodson (2014) suggest that teachers often try unsuccessfully to achieve learning about science content, the nature of science and science investigation processes all through one investigation. While these teachers also had multiple goals for science investigation, they suggested a major purpose for including it was the opportunities it provided for building an understanding of how science happens

—observing, inferring, talking about ideas and suggesting possible explanations, finding the same thing in different ways, and comparing methods.

Both believed the teacher has a role in guiding and supporting students to investigate scientifically, helping them to observe carefully, and learn to record in detail and methodically. They thought more scaffolding and guidance were needed for the younger ones, but students could still be encouraged to wonder and develop questions and to investigate to find answers. They anticipated the kinds of things students could learn in a given context, but they were flexible:

> I know that I want some tally charts for maths and things like that, so you know, I'm going to have something like that but some of them will draw. Some of them will talk. Some of them will write a word … I'll know the connections I want to bring up and they'll say, ooh yes, we did that in maths or we do that when we count to a hundred but often you just need to … put it in their heads. (Alison)

In summary, student questions and ownership and authenticity were seen as important components of scientific investigation at school, but the teacher had a major role in enabling learning and helping students to make connections.

3.4   Teachers' Practice of Science Investigation

Three lessons that included science investigation were observed in each teacher's class. A summary of each teacher's practice is described here. As in most New Zealand primary schools, science in both classes was not regularly timetabled on a weekly basis; instead it was planned as a unit of work, i.e. a period in the term during which science was a major focus and taught two to four times a week.

The findings reported here were developed from: audio and video recordings of the three lessons; completion of a guided observation schedule and running observational record for each lesson; conversations with the students as they engaged in practical tasks; examples of student work; and teacher resources such as unit plans and resources they had developed themselves. The data collection and analysis are described more fully in Chap. 2. A summary of observations from Alison's lesson sequence is provided in Appendix 3.1 as an example of the way in which each teacher's practice was analysed. We begin with a summary description of the three lessons observed in each classroom.

Science Investigation in Alison's New Entrant Classroom

The unit was entitled 'Bubbles'. The idea for the unit and some of the activities came from a primary science resource—science postcards—developed by New Zealand primary science educators which link a science investigation to a picture book (www.sciencepostcards.com). Prior to the observed lessons, the students had listened to the story *The King's Bubbles* (Paul, 2007) and gone outside to blow some bubbles. Alison had taken photos and used them to develop a class story on large sheets strung across the classroom. The teacher and students had together developed a 'what we know about bubbles' chart. Although sequential, there was a gap of over a week between the first and second observed lessons as the class was involved in other school activities for that week. The second and third observations were on consecutive days.

Alison was well-organised for each lesson—equipment was prepared and ready for use. She appreciated the needs of her students, anticipating and resourcing potential directions. For example, in this unit she thought students may want to investigate if different shapes of the bubble blowing equipment affect the shape of the bubble, so had collected an assortment of different shapes and wire for blowing bubbles. One of her students was an advanced reader so Alison had given her a school journal[2] article containing a bubble mixture recipe that the student had made up ready to test. Alison has developed strategies for managing student practical work with the young students, for example, she recognised that they mostly work individually so there was enough equipment in a carry-out tray for each student to have their own bubble blowing loop and mixture: 'They will all need their own— not so good at sharing at this age!' Spare loops and mixtures were available in case of breakages and spillages. A bucket of water was ready in case bubble mixture went in their eyes. The suggested learning outcomes Alison had recorded in the unit plan were that students would be able to:

- Blow bubbles in different ways.
- Describe some properties of bubbles, e.g. what bubbles are made of and colour/ shape/weight/composition.
- Share their ideas about how bubbles form.
- (Extension): Explain which bubble mix (bought or homemade) makes the best bubbles—biggest/longest lasting.

Lesson One: Developing Student Questions

The focus of the first lesson observed was for students to make and talk about their own observations of bubbles and to develop some questions that could be investigated in the second lesson. Alison began by re-reading the original story, which raised a number of ideas that had the potential for investigation. She then read out the ideas each student had shared in the previous lesson about their bubble experiences: 'Steven said that bubbles pop and they never come back, Laura thought

---

[2]The School Journal is an instructional reading resource published at three different levels three times a year and freely available to teachers in New Zealand schools (http://literacyonline.tki.org. nz/Literacy-Online/Planning-for-my-students-needs/Instructional-Series/School-Journal).

they could look blue, Katy got a bubble pipe for her birthday, Daniel said "They float—I don't know why!"' Alison explained they would be going outside to investigate bubbles: 'Blowing bubbles is a really good way to learn about them because we can look and observe with our eyes and see what's happening.' She emphasised some starting points for observation that arose from the book:

> There was an interesting question in the book: it said that there were puddles on the ground. So, if we go out and blow bubbles I want you to look and see if there are puddles on the ground ... I wonder if there will be?

She led the students through an imaginary observation of blowing bubbles, watching to see if they popped and looking on the ground to see if there was anything on the ground. 'Ooh look ... is there something on the ground?' Students modelled blowing bubbles using the plastic loop, and Alison explained what to do if they got bubble mixture in their eyes before going outside.

Outside students were each given a bottle of bubble mixture and a plastic loop. There were multiple reminders about useful observations: 'We are looking to see if they go up or down or if they pop ... You can see what happens if you blow hard or if you blow softly – I wonder if it makes a difference? ... Look at the ground – what can you see on the ground?' Although Alison endeavoured to guide and focus students' observations in this way, they essentially followed their own line of play or investigation—it was hard to differentiate between the two, as will be discussed later. After about 20 minutes, the students returned inside to the mat and Alison invited them to share their wonderings or questions. Shared wonderings were recorded with the student's name. They related to students' experiences and were not necessarily observation focused: 'Can you catch a bubble on your bubble stick and use it to blow another bubble without using more mixture?' Alison prompted others that could lead to productive investigations and learning: 'Barrie I know you noticed something about colours—did you have a wondering about that?' She also used this reflection time to draw on students' experiences to confirm that bubbles left a puddle when they burst, a gentle nudge towards the idea that things do not simply disappear, conservation of matter being an important substantive concept for students to develop over time (Fig. 3.1).

Lesson Two: Observing to Find Answers

In the second lesson, the class investigated three of the questions identified in the previous session: Do different shaped blowers make different shaped bubbles? Can bubbles be caught on fingers? What happens when bubbles pop? The student who had raised the question read it out and the other students were invited to think about how it could be answered: 'What could we do to find out?' However, the investigations were strongly teacher guided and designed; Alison provided appropriate equipment and suggested relevant observations: 'I wonder what sort of bubbles this shape will blow. Do you think they will be the same as the bubbles we blow with this one? Let's find out!' Students worked individually with the equipment

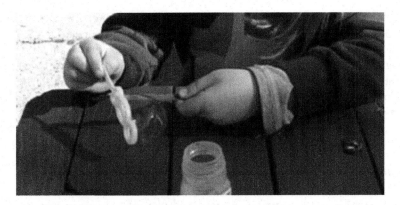

Fig. 3.1 Children sometimes spent considerable time pursuing their own exploration. This child tried repeatedly for several minutes to see if a bubble could be caught on a leaf

provided and, as before, tended to follow their own line of investigation at different points, but sometimes, with direction, exploring the investigation question. Sometimes Alison would ask a student to explore an aspect, such as whether a different shaped bubble blower blew a different shaped bubble, while everyone else watched. In this way, she focused them all on the same observation at the same time. Again, Alison finished the session with the students returning indoors for a reflection time. She asked questions that connected students' thinking back to the investigation questions, helping them to remember observations they had made to answer them. Most questions were easily answered in this way. Observations were generally conclusive; for example, they found that bubbles could be caught on fingers, and different sized bubbles could be formed, but they were all round and the shape did not change with different shaped bubble blowing devices. Alison suggested that another investigation question about what happened when bubbles popped might be able to be answered by looking in a book, although many students had responded that they could look at bubbles to find out the answer. This referral to books as a warrant for belief is quite common among primary teachers (e.g., Appleton, 2006); however, as we shall see next, Alison worked hard to address this quite abstract concept in a practical way using a model.

Lesson Three: Investigating Models to Develop Understanding

In the final lesson, in the sequence, the class investigated the last two student questions. Alison shared information from a book describing what happens when a bubble pops. She used a balloon brought by a child to illustrate the ideas raised in the book; the child blew up the balloon and helped all the others to feel that it had air inside and they explored the idea of 'skin' raised in the book. Alison then used a demonstration with student observation of bursting bubbles, bursting balloons to make comparisons between the skin of the balloon and the skin of the bubble to develop students' ideas about what bubbles are made of, and what happens when they pop. The second part of the lesson explored the final student question about the colours seen on bubbles. The class went outside and observed bubbles in the sunlight. Alison's focus was on observing colours and associating them with light: 'What colours can you see? … I want you to look and see what happens in the light. … I'm looking for colour words—call out colour words you can see.' Students returned indoors to draw a bubble and add the colours they had seen.

Science Investigation in Patsy's Year 5/6 Classroom

Patsy's unit was entitled 'Weird and Wonderful'; it was a unit she had used previously and felt students had enjoyed it, but also that it would culminate well in a planned cross-class event, *The Mad Hatter's Tea Party*, later in the term. The three lessons observed were those that formed the unit. Her plan listed the Nature of Science strand achievement objectives from the Understanding about Science and Investigating in Science strands of the New Zealand curriculum (Ministry of Education, 2007). Words were underlined, showing the emphasis for this unit: 'Scientists work together and provide evidence for their ideas. … Ask questions, find evidence, and carry out investigations to develop simple explanations.' A substantive concept about physical and chemical properties was also included in the plan, presumably to support teachers with identifying and supporting the development of these concepts. Key words were again emphasised:

> Substances have chemical properties that can be used to group them. Chemical properties involve reactions with other substances. New substances are formed as the reacting chemicals bond together in a new way. The chemical properties of a substance determine the chemical changes it can undergo … We investigate chemical properties of a substance by observing the way substances react with certain other substances.

As proposed by Hodson (2014) and Abrahams and Millar (2008), there appeared to be multiple objectives for the lesson sequence. In hindsight, it seems Patsy's purpose appeared to be for students to learn to make detailed observations and to associate observational rigour with science, as these things were often emphasised. A supporting goal appeared to be to provide her students with experience of a range of chemical reactions.

The activities mostly came from the resource Making Better Sense of the Material World (Ministry of Education, 2001) provided free to schools to support the previous curriculum. Like Alison, Patsy had organised all the materials herself, going out at lunchtime for some last-minute purchases. All three lessons were observed over the space of a week and, as in Alison's class, occupied the whole afternoon session, except for the last one which took the whole session between morning tea and lunch.

Lesson One: Introducing Acid-Carbonate Reactions; Identifying Acids and Carbonates

This lesson introduced the topic of chemical reactions and began with teacher talk and questioning about students' current understanding of the topic. Patsy linked back to some work with clay that students had done earlier in the term. Their ideas about chemistry were collected on the board; Patsy used a library book about elements to explain that everything is made of chemicals. When asked about acids, students' examples included lemons and sourness; Patsy also checked students' understandings of liquid and solid and linked this to prior experiences of making baking soda and vinegar volcanoes to remind them of what a reaction is. She shared the learning intention with students: 'We are going to do two things: we are going to find out about acids and carbonates, but we are also going to think about working like scientists—if you are a scientist working out an acid from a carbonate, how do they do that?'

To begin the practical session, Patsy gathered the class round a table and demonstrated mixing baking soda and vinegar. She emphasised ways to make careful observations: 'watch carefully' and asked them to decide what to watch:

> Reactions might not be huge – you will need to be really focused … Could we measure the time it takes to bubble? … Make sure you record your observations – not what you *think* happened, but what you observe. If you have questions or ideas you can record them too – we might investigate them later.

The purpose of the demonstration was for students to experience a reaction and know what to look for in an acid-carbonate reaction: 'So, you all know what a reaction is. When you go off to do your own you will have to look very closely; what might you be looking out for? Smell? Maybe, but the main thing—Bubbles … Bubbles.'

For the second part of the lesson, Patsy had organised three tables, one with known acids (largely fruit juices), one with known carbonates (washing soda; soap flakes; chalk; crushed eggshell), and a third table of 'unknowns.' Students were provided with a recording sheet, paper plate and a teaspoon. Patsy asked them to first mix some known acids and carbonates and record their observations. She noticed during this time that students were often not observing the bubbles as they were not waiting long enough. She emphasised careful and prolonged looking as an important part of being a scientist: '…really look at it, get down and level with it and have a good old look!' She also put out magnifying glasses which the students enjoyed using to see if there were tiny bubbles as evidence of a reaction. After a brief think pair share about investigating unknowns, students were given a plate divided in two with a side for testing with a known acid and a side for testing with a known carbonate to use in identifying some unknowns, again with a reminder to 'wait and watch. Emily's chalk bubbled just as she was about to throw it away!'

Patsy moved around working with individuals, encouraging them to observe carefully and record details of what they did and what resulted. As with the younger students in Alison's class, this investigation, although initially teacher directed, turned into student-led exploration. There was enough equipment for students to explore individually for themselves and they made the most of the opportunity to observe what happened when they mixed different combinations. Although many of these explorations were far from orderly examinations of different combinations, students took seriously their teacher's instructions to record exactly what they did in detail, like scientists, as can be seen in Fig. 3.2.

Many students recorded careful and detailed observations, even though the recording sheet did not strongly scaffold orderly investigations or detailed recordings (see Fig. 3.3).

An orderly clean-up routine meant that materials and tables were tidied and equipment cleaned and put away. Patsy then asked students to write down how they knew their unknown substance was an acid or a carbonate, helping them to link to observations and evidence when making inferences.

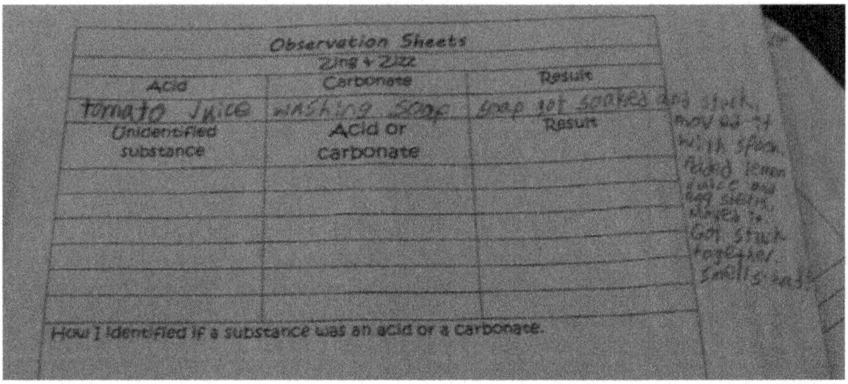

Fig. 3.2 A student's detailed recording of what they did in their investigation

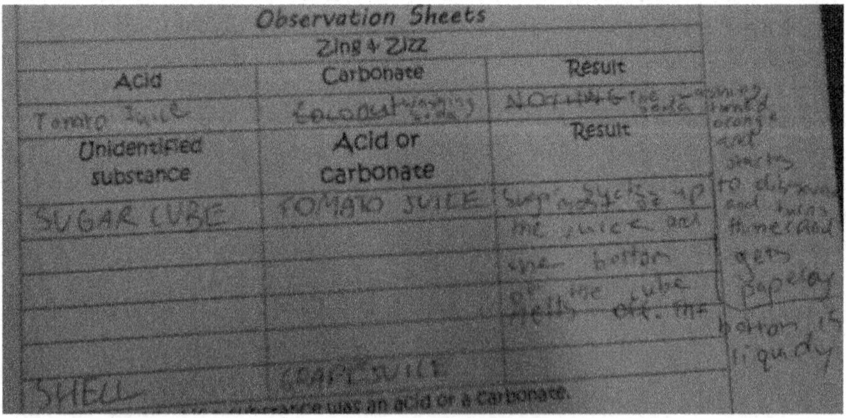

Fig. 3.3 Example of a student's recording of detailed observations

Lesson Two: Making and Investigating Sherbet

Like Lesson one, Lesson two began with teacher talk and questioning, this time reminding students about acid-carbonate reactions observed in the previous lesson. There were two investigations in this session. The first was a structured sequence in small groups working through each stage at the same time to make sherbet. Students used a recording sheet to observe what they did and what happened at each stage. Patsy emphasised aspects of science throughout: 'usually in science we don't taste things, but today we are; … often in science we repeat things; … 'part of science is measuring out things carefully so, it might take a little bit longer; we are all going to do that together.' Students worked in small groups with a student designated to collect the group's ingredients. Patsy asked students to describe the taste at each stage: 'Not what it is, but what it's *like*', maintaining her focus on detailed observation, this time using a different sense, although the focus on visual observation also continued: 'Watch it when you mix it on the plate—does it change then?'

Working in a stepwise and orderly manner all together, but with an individual set of ingredients, students tasted each ingredient and the effect of adding it to the mixture. They carefully followed instructions as Patsy asked them to taste, mix, share ideas and record their taste observations. She checked what the students noticed after each addition. After the last stage, when the acids were added to the mixture containing the baking soda, Patsy asked the students to record why they thought it foamed. They had different ideas about these observations, most connecting them to their previous experience. Some explanations proposed were rather circular: 'it happened because of the racshon [reaction]'; some were confused, understandably as it was not discussed which ingredient may have been the carbonate: 'This happened because of the citric acid and the tartaric acid fizzing.' At least two students recognised the significance of the liquid saliva (see Fig. 3.4a, b); however, they may well have been thinking that the liquid was producing the reaction. As can be seen, students provided observations as explanations.

One student showed the researcher that 'When you have a wet finger you can see it making foam' (Fig. 3.4). His explanation was that the water 'mixes with the acid and makes it liquid and then mixes with the carbonate and makes it foam,' a reasonably accurate step on the way to the scientific explanation.

In the second investigation, about making fizzy drinks, lemon juice was substituted for the solid acids and green food colouring added. Patsy demonstrated what happened initially. Students then had to decide in their groups which of the ingredients—icing sugar, baking soda, or lemon juice—to increase and then observe the effect. Patsy emphasised prediction: 'Talk together about what effect that is going to have—write down your prediction.' Students discussed the choice seriously in their groups: 'If you add more sugar you'll taste it more' … if we add that [baking soda] it might fizz more.' The degree of change was managed by Patsy; once the group had decided what they wanted to change, she gave them double the amount of the original recipe. She checked that across the groups all options were being investigated. Each group first tasted a sample of the original mixture, then

(a)

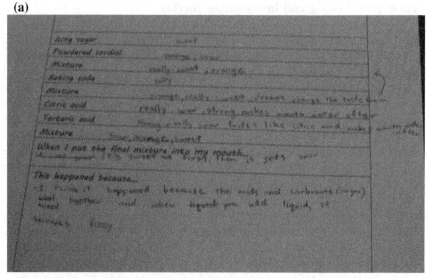

(b)

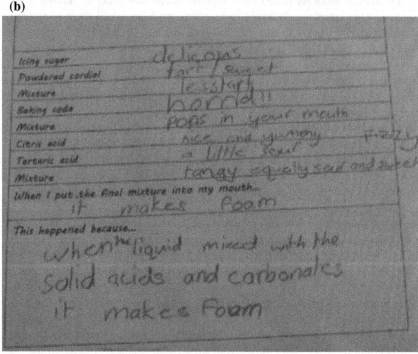

Fig. 3.4 These examples of student work show understanding of the need for a liquid (**a**); both students provided observations rather than possible explanations (**b**)

Fig. 3.5 A child studies
carefully what happened to
the mixture when his finger
was wet

compared their changed version. Observation and recording a detailed description
continued to be emphasised: 'You should all have something written down about
the taste, the fizz and the flavour of Number 1 – put that in the box now … Not
whether you like it or what it was about it that meant you didn't like it—what was it
actually like?' As time was short, once they had completed the mixture with their
increased ingredient, students completed their observation sheets about taste, fizz
and flavour and then tidied up for the day.

Lesson Three: Exploring a Range of Reactions

In this lesson, students in small groups explored a range of reactions and physical
properties in several activities: Gloop (PVA glue and boric acid), Oobleck (corn-
flour and water), Billowing balloons (filling a plastic glove with the gas released
from a baking soda and vinegar reaction), Catastrophic Currants (currants floating
to the surface with bubbles of gas from a baking soda and vinegar reaction) and
Elephant Toothpaste (foam produced by the hydrogen peroxide, detergent and an
active yeast mixture). The activities came from *Making Better Sense of the Material
World* (Ministry of Education, 2001). To begin the session, Patsy told students they
were required to be '"observing" scientists and "recording" scientists and "thinking
about what's happening" scientists… This session is about observing and looking
and touching and feeling and recording what you noticed.' Patsy set up each
investigation as she demonstrated what to do and look for; the class discussed their
initial observations and ideas then recorded what they had seen. Students then
circulated through each activity in groups, with time given to making sure they had
recorded observations after each round. They used a single piece of paper folded
into six sections for this. Students were highly engaged both in watching as each
activity was prepared and in exploring each mixture. Patsy commented at the last
round:

Fig. 3.6 Examples of students' detailed recording of observations

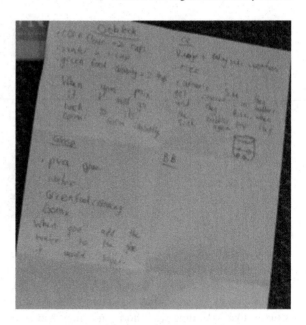

> I know you've been doing this for a while, but it's really important. Scientists often have to keep observing and keep the same kind of observations going even though they are really tired of it, because they may miss something. So for this last one really focus on observing carefully.

Students played with and explored each activity, and took the careful recording instruction seriously during the time allocated for it, as shown in Fig. 3.6.

The lesson concluded with a tidy-up time.

Science Investigation: Teacher Beliefs in Practice

The descriptions provided above show a strong degree of congruence between the beliefs the teachers espoused about school science investigation and their classroom practice.

Both primary teachers strongly believed that science investigations were an opportunity to reflect on importance placed on observation and evidence as a significant feature of science, and both provided a wealth of practical investigations, including open-ended investigations to which they themselves did not know the outcome. They explicitly linked practical activity and investigation with 'doing science' or 'being scientists.' The teachers selected the context for the unit and largely the investigations that were done, although Alison chose investigations that addressed students' wonderings and questions. Both chose contexts in accordance to their beliefs that science investigations should be on topics of interest and relevance for their students. Alison chose 'Bubbles' because of the engagement and curiosity showed by her students after a practical bubble blowing reading response activity. Patsy chose her 'Weird and Wonderful' unit on reactions because of her previous students' enjoyment and engagement with it. Their knowledge and beliefs about students' interests appeared well founded, as observations of student engagement levels showed most students were enthusiastic, attentive and on task for all observed lessons, although perseverance was less consistent than the other observed engagement aspects.

Student ownership was a feature of science investigation highlighted by both teachers, but particularly Alison, and a characteristic identified by Millar (2010). In Alison's class, as may perhaps be expected for 5-year olds, there was a high degree of teacher input in selecting and guiding investigations; however, a high degree of student ownership was also enabled at this new entrant level. After the initial free exploration of bubbles, the follow-up investigations all resulted from student questions. In Patsy's class, there was less emphasis on wonderings and questions that students developed, although she did suggest that students explore their wonderings about a given investigation and provided time for personal exploration in Lessons one and three.

A strong emphasis present in Patsy's beliefs about investigation was that students should have the opportunity to collaborate. In her class, students worked mainly in small groups with a group set of materials; as previously stated there were also opportunities in Lesson one and the first part of Lesson two where time and enough equipment were allowed for individual investigation. In contrast, Alison's younger students investigated individually; she explained that she knew from experience that each student would struggle to share equipment and would want to follow their own ideas, and so provided enough equipment for individual investigations.

A second, and very strong, emphasis in Patsy's beliefs about the nature of school science investigation was that students should be clear about how and what to observe. Again this aspect featured clearly in her practice, as evidenced by many of her comments described above. Students were reminded often to decide exactly what they would observe, to observe closely, and follow it carefully, which senses to use and to persevere with their observations. Student examples of the kind of detail that could be recorded were shared, for example, about ways to describe different tastes, and time was provided to record detailed observations. Alison, too, focused frequently on careful observation and helped students use their observations to answer their questions. Both teachers identified careful observation explicitly as a scientific behaviour.

Both teachers also agreed that science investigation should be purposeful. Each appeared to have a range of purposes in mind for the investigations in the observed lessons, including development of some substantive understanding, as will be discussed shortly, but for both teachers it appears that building students' under-standing of what it means to do science and be a scientist was prioritised over development of substantive concepts. Little time was devoted in Patsy's class to developing concepts about reactions, acids or carbonates, although the investiga-tions she included could naturally have led to that outcome. Alison spent most of her time on investigations that answered student questions, although one of these involved substantive concepts about the structure of bubbles. Both teachers talked with their students frequently about behaving like scientists and what scientists do. Alison referred often to a poster of 'The Parts of a Scientist' displayed at the front of her classroom: 'Let's put on our scientists' eyes ... Ailsa, you have a curious mind like a scientist.' This emphasis may well have been as a result of their experience as Primary Science Teacher Fellows, working alongside scientists, and is probably not typical of most New Zealand primary teachers. An aim of this programme is to raise the profile of science in teachers' classes and communities (Anderson, 2013).

Congruence between espoused beliefs about science teaching and learning has been demonstrated previously among New Zealand primary teachers (Anderson, 2015). Strong connections between classroom practice and teacher beliefs about science teaching and learning have also been noted in Australia (Fitzgerald, Dawson, & Hackling, 2013). In New Zealand, such alignment may be a result of a broad national curriculum that expects schools and teachers to respond to local needs and contexts. The New Zealand curriculum provides latitude to select, interpret and implement the different learning areas in many ways. Such freedom also means that teachers' beliefs about the nature of a subject, which content is important to teach and learn, and indeed even whether science should be taught at all in any given year, may hold greater sway than in countries with more prescribed curricula. In this case study, the teachers' beliefs and practice align with the direction of the New Zealand curriculum, which places the Nature of Science strand as overarching and compulsory (Ministry of Education, 2007); however, there is much evidence suggesting that such practice is not common and that the Nature of Science strand is not commonly addressed in New Zealand primary schools (e.g., Chamberlain & Caygill, 2012; Education Review Office, 2012).

3.5 Teacher Strategies that Provided Opportunities for Learning Through Investigation

Although the major purpose of this phase of the study was to gather baseline data, we noticed teacher practices and strategies that provided useful opportunities for learning through science investigation, several of which have been alluded to already but which are worthy of teasing out in more detail here.

Supporting Students to Investigate

Range of Approaches to Investigation

As part of the study, we considered the range of investigation types that teachers introduced to students. As described in Chap. 2, an aim of the New Zealand curriculum's Investigating in Science sub-strand is that students experience a range of investigation approaches: classifying and identifying, pattern seeking, exploring, investigating models, fair testing, making things or developing systems. We were curious to see if students experienced particular approaches in primary and secondary school, or if there was an overall dominance of one type, such as fair testing. The investigations observed in the two primary classrooms at this time were nearly all explorations—investigating what something is like, or whether something happens in a given circumstance. Classification and identification were used in Patsy's first lesson. Students were provided with experiences of the way known acids and carbonates behaved, which they then used to classify unknown household substances. In Patsy's class, students investigated reactions by making sherbet, identifying which substances led to a reaction occurring. They also made a green fizzy drink, investigating the impact of increasing one ingredient—an introduction to fair testing. Alison investigated a model with her students, using a balloon to help them consider how bubbles were structured.

Time and Equipment for Play/Exploration

A range of investigation types was observed, but an exploration of each context formed a major part of students' experience of science investigation in both classes. Provision of *time* and *equipment* were critical in enabling students to play with and explore the context. Alison used this time to model wondering and prompt students to test out ideas: 'You can see what happens if you blow hard or if you blow softly —I wonder if it makes a difference?' However, children rarely explored in the directions suggested for very long. In both groups, it was impossible to tell when investigation became play and vice versa; one rapidly became the other, but at various times many students spent time quietly intent on their own line of practical inquiry. In Alison's class, students chased the bubbles as they blew in the wind, they tried to see if they could join bubbles together, how many they could blow in one go, and what happened if you blew into the mixture rather than through the loop. One question did not need to be answered before another took precedence. Attention during their own play/investigation could be sustained for long periods of time, showing great perseverance. For instance, one student spent 10 minutes repeatedly trying to transfer bubbles from her loop to a leaf (see Fig. 3.1). However, it could also be short-lived and fleeting as another student tried once to see if two bubbles could be joined together then, distracted, chased across the playground as the wind caught one. Similar behaviours were observed among the older primary students during exploratory times in Patsy's class; students would add more of one substance, or a range of substances one after the other, to another, just to see what would happen, or would spend minutes watching the currants stop dancing, and add more baking soda and watch it all over again. The equipment impacted on the play that occurred; the children played and explored to the extent of what was available. Both teachers provided plenty of equipment which facilitated the group and individual exploration.

Encouraging Wonder and Developing Investigable Student-Owned Questions

The initial opportunity to play and explore was important for Alison in developing student-led investigations. She modelled wondering: 'I wonder why it sparkles in the sun?' and deliberately noticed students' wonderings during this time. She actively supported them to wonder during the class reflection time following their outdoor play/exploration: 'Sit down and close your eyes. Picture outside with the bubbles and think about your wonderings—what do you wonder?' She helped them turn wonderings into questions: 'Is that a question? … How can we turn that into a question?' She recorded each wondering as a question and displayed it along with the child's name, using sticky notes on a poster. Alison made a point of acknowledging student ownership of questions: 'Aggie, come and read out your question … Barrie, what answer have we found to your question? … Today we are answering Thomas's question.' The student concerned had their question and name written on a card and, with teacher help if necessary, read out their question before the class carried out the practical investigation to answer it.

Teacher Scaffolding to Build Rigour in Investigating

Both teachers intervened to make investigations more orderly or valid. Alison provided equipment for the class that would enable answering of students' investigation questions, for example, providing a range of different shapes for them to blow bubbles with. She guided students to make relevant observations for each question: 'Does the square shape make square bubbles?' She spent time with individuals, listening and focusing their excited noticing, or getting them to explain what they were looking for during exploration and play: 'How many bubbles have you got? Are they all round? … Oh, it does sparkle in the sun! What colours can you see?'

Patsy noticed when students were adding multiple substances in the first lesson, and suggested they note down what they had added and what happened as they added each one. She did not curb their enthusiasm and exploration by making them systematically try just one thing, but she did signal that noting what they had done and its effect were important in science. In the green fizzy drink investigation, in Lesson two, students chose which ingredient they wished to observe the effect of, but it was Patsy who restricted them to changing the amount of just one substance and calculated how much more they should add, doubling the amount in proportion to the recipe.

Teaching Students to Observe

As has been noted throughout this chapter, both teachers supported their students to build their observational skills. They prompted and questioned as already described, but also used teacher modelling. For example, when wanting students to observe the colours of bubbles in the light, Alison blew some bubbles in the sunlight and watched one bubble closely using a talk-aloud: 'I'm blowing a bubble and I'm following that one with my eye.' Patsy often demonstrated an activity before the students did it and would point out the features they could observe. 'Now as I put in the baking soda, choose one thing to watch carefully, maybe the liquid, or the baking soda, and keep watching it. Watch what happens to it.'

Patsy pressed for detailed descriptions: 'Saying 'yuk' is not a scientific description; saying whether you like it or not, is saying a different thing to describing what it's like.' She publicised students' descriptive words as examples for others: 'So, it tastes salty … sour …' Alison, in the class reflection time that followed every investigation, encouraged students to describe what they had seen. In the final lesson, she focused quite heavily on students naming the colours they had seen: 'Some of them don't talk a lot and I want them to name the colours.' She also provided an outline of a balloon that they could colour with the colours they had seen.

Opportunities to Develop Substantive Understanding

As indicated earlier, both teachers included substantive goals for learning in their planning, although developing these goals did not appear to be as much of a priority as providing experiences of the evidence-based nature of science and making scientific practices explicit. Three strategies were noted that supported students to make connections to big ideas.

Focusing Students on Key Observations

As described in Lesson one, Alison wanted students to understand that something was left after a bubble popped, it did not just disappear. Before going outside, she made suggestions about what they could observe: 'We are going to be scientists and use our eyes to observe … we are looking to see if they go up, or down, or if they pop … is there anything on the ground where they pop?' She revisited this idea in the reflection time after the investigation: '…and we saw that there were puddles on the ground when bubbles burst.' These observations contribute to the understanding of what bubbles were made of, and also build towards the larger science concept of conservation of matter.

Similarly, in Lesson two, Patsy focused students' attention on key observations. She strongly structured students' participation in the sherbet investigation. To help them identify the mix of substances that creates the fizzing in sherbet, she made sure all the students carried out each step of the investigation at the same time, recording their observations after adding each ingredient.

Connecting Ideas and Observations

As described in her lesson sequence, Alison used information from a book, illustrated by using a model of a balloon, to identify that bubbles had a soapy skin, a bit like the balloon's skin, and were filled with air. A child blew up a balloon and Alison ensured each child had an opportunity to feel the skin and that it had air inside. She then physically blew bubbles with the students to point out connections between the balloon model and the bubble structure that they were trying to understand. As they blew bubbles, Alison talked about the balloon being made of a soapy water skin, filled with air, and that it bursts when the skin dries out and thins and the air escapes just like 'when we pop the balloon and the soapy skin falls down like the broken balloon, we see the puddle.'

Although learning to make observations appeared a more imperative goal than learning from them, Patsy would sometimes connect to substantive concepts:

Patsy	What are we looking for?
Sylvia	Bubbles
Patsy	Why do we think we might get bubbles?
Steven	Because there is a reaction.

The recording sheet Patsy used in Lesson two also required students to consider why they thought the sherbet foamed, demonstrating another way in which students could be guided to consider explanatory ideas.

Shaping Class Discourse

Where Alison was drawing on students' observations, either to answer the investigation question as in Lesson two, or develop some conceptual ideas as in Lesson three, questioning became more directive and specific. Questions were quite closed and directive when she wanted to draw conclusions: 'Who caught a bubble on their finger? So can we catch bubbles on our fingers?' Students chorused 'Yes!'. Closed and leading questions were used for more inferential and abstract questions to force the response she wanted:

Alison	What makes bubbles float?
Several students	Wind!
Alison	Wind and ... what do you blow into the bubble? What do you breathe out?
Several students	Breath.
Alison	And what's in your breath? Ai...ai...? (sounding out)
Students	Air!!

Less productive answers were often not given too much attention, and more con-ceptually productive answers were celebrated and publicised. The following excerpt illustrates this point. It occurred in Lesson three after students had listened to the information about bubbles from a book, then examined balloons where Alison had asked them to feel the 'skin' of a balloon. Students then observed her blow a 'soap balloon' which 'we call a bubble' and watched it pop:

Alison	What do you think broke?
Sally	The bubble ... the bubble.
Alison	Does not respond.
Steven	The dry skin.
Alison	The dry skin! Good boy Steven! That's right, if the skin on the balloon gets dry it pops!

This use of directed questioning was used by Alison to steer the discussion in productive directions for conceptual development, and she then paraphrased the correct idea to validate it with her teacher authority and make it public.

Such management of classroom discourse has been observed by other researchers. Treagust (2007) found teacher questioning and intervention played a significant role in the quality of discourse and ways in which students learned and understood science. Mortimer and Scott (2003) classified the communicative approaches used by a group of teachers with students in secondary science. They noted that a teacher's ability to manage these discourse forms helped in identifying and supporting the development of students' substantive understanding. We can see Alison manipulating the discourse in a similar manner in the above examples, using what Mortimer and Scott would call an interactive/authoritative communicative approach where the teacher builds a high degree of interaction but despite this, little attention is paid in reality to student ideas; they are ignored or discounted as the aim is to reach one specific point of view. Alison elicited ideas, used more closed questioning to hone in on a particular idea, and then ignored responses that did not fit with the scientific concept, celebrating those that did. At other times, for example, when developing questions to investigate, she used a more interactive dialogic approach, where she created many opportunities for interaction, but this time multiple ideas were elicited and taken account of even though they differed from those Alison was keen for students to investigate.

**Making the Nature of Science Explicit and Developing Science
Capabilities for Citizenship**

As we have seen throughout this chapter, both teachers constantly linked the
practical investigations and careful observations the students were making with the
discipline of science. They positioned their students as scientists and labelled the
work they were doing as science: 'What a scientist you are!' and 'You ARE a good
scientist, Katy!' Alison commented during the exploration/play session. They also
talked about scientists and described the ways they work. Patsy told students that
'quite often scientists collect their data, they go out and investigate; they make sure
they have recorded their data carefully, and they go away and think about it later—
the whys and the wherefores.' Similarly, Patsy spent time with individuals, asking
what they were doing and observing, and encouraging them to persevere with, and
record observations: 'Note down what you did, and carefully record what happened, like scientists.'

As part of the study, we examined teachers' practice for development of the
Science Capabilities for Citizenship (Bull, 2015), a functional means of addressing
the Nature of Science strand advocated by the New Zealand Ministry of Education
teacher support website for science[3] (see Table 2.1). Although these capabilities
were not explicitly a focus in either teacher's planning, both of them included many
opportunities for learning to gather and interpret data—the first of the five capabilities. Developing this capability involves learning to observe and make inferences based on observations, separating inference from observation. The emphasis
on supporting students to observe carefully and guiding their observations has
already been noted many times. Alison also provided opportunities for students to
make inferences and use evidence to support ideas, another of the five capabilities.
Her comments showed that inferring from observations was a valued skill. For
instance, when Simon, who had been trying to join two bubbles, said 'It got bigger
because one bubble went inside another one,' Alison commented 'Good explanation!!' She used class reflection time following the investigations in the second and
third lessons to help students use the observations they had made to formalise an
answer to the question that prompted the investigation:

> Alison   Think about what we saw- who caught a bubble on their finger? ... Can we
>   catch bubbles on our fingers?
> Students   Yes!! (chorus)

Many of the student investigation questions, as in this case, did not require inferences to be made; answers could be directly observed. The way she helped students
use their observations to justify their answers to the investigation questions also
suggests that Alison wanted students to understand that scientific knowledge is

---

[3](http://scienceonline.tki.org.nz/Science-capabilities-for-citizenship).

based on observations of the natural world. The focus on using evidence to support ideas was common and occurred spontaneously throughout lessons:

Alison   How do you know that it's got air inside?
Barrie   Cos it comes out and you can feel it

Alison's use of the model to represent a bubble provided students with an experience of the capability 'using and interpreting representations', which includes working with models and diagrams. In Patsy's class, students spontaneously included diagrams in some of their observations, but this was not a focus for learning. The other capabilities—'critiquing evidence', and 'engaging with science', in which students use all the other capabilities to engage with a science-related issue—were not observed in either class.

3.6 Student Learning from Investigation

In this section, to examine the learning that occurred through these investigations, we look across researcher classroom observations, examples of student work, a whole-class post-unit questionnaire (Patsy's class only, see Appendix 3.2), interviews after each session with individuals (Alison's class), and a focus group of six students (Patsy's class).

Some of the learning Alison intended, i.e. observing bubbles and suggesting how they form was established by some students, but not all. It is not clear how students' ability to observe progressed. They tended to observe what they wanted to find out, rather than aspects the teacher suggested they focus on except when directly asked individually, or as in the last lesson where all students appeared focused on the colours through continuous teacher prompting and where the colours were very obvious. Students could all chorus in response to Alison's question that bubbles could be caught on fingers, and that they made the same shapes, and several students noted during class reflection that some bubbles were bigger than others but were still round. They were less clear about what made bubbles float; although the answer 'air' was attained from the class, it was through strong teacher prompting, and it was not evident that they understood what it meant. When asked about their learning, students were inclined to talk about what they noticed and did individually. In interviews, they described their learning as the aspect they were interested in or pursued during the lesson. For instance, after Lesson two where the teacher's focus was on bubble shape and catching bubbles, one student's focus was on making lots of bubbles: 'I blowed in the cup with the bubble mixture and then it made a lot of bubbles and when I blew it more, more bubble came up. I blew some more and then some of them popped.' What they investigated or achieved for themselves during their play/exploration was of more importance and relevance than the aspects that Alison focused on. Students appeared to take as the most significant learning from their investigations the things that they personally investigated.

In summary, there appeared to be two dimensions to substantive learning for students in Alison's class in the observed unit. They experienced a very structured sensemaking during reflection times that they complied with. Students tended to try to establish what it was the teacher was looking for and gave her the answers she wanted in these types of discussions. It was not clear how meaningful this was for each student. The other dimension was the learning described above that occurred as they followed their own lines of inquiry. This learning was mostly descriptive in nature and appears to have held more personal significance for students. In terms of investigative skills, with strong teacher support, students made wonderings that the teacher turned into investigable questions. Students mostly used unstructured play with the equipment provided to follow their own line of investigation. When equipment was specifically selected by the teacher, these investigations more closely followed a focused line of inquiry. Most students appeared to be able to make simple observations when these were obvious and specifically directed by the teacher. Either from their own or teacher supported investigations most students could use observations directly to answer, with teacher support, investigation questions that were immediately obvious. With regard to learning about science itself, it became apparent through interviews with students that although Alison frequently referred to science, the students described their learning in terms of what they had learnt about bubbles. They were confused when asked what they had learned in science. It was hard to probe what they had learned to do that they identified as scientific.

For Patsy's class, students interviewed after the sessions similarly described things they did as their learning: 'If you mix flour, vitamin C tablets and sugar cubes together, it turns all gloopy and rough.' Probing of what they took from lessons revealed general understanding of observations that would signal a chemical reaction: 'We were looking for fizzing, fizzing … or colour change … forming a new type of, sort of like multiple things … Two things make another thing.' Some students appeared to think that all substances had to be either an acid or a carbonate, or that dissolving, as in a sugar cube dissolving in lemon juice, constituted a reaction: 'I got the sugar cube. I got the last of the lemon juice and by the time I finished squeezing it, there was a big hole in the sugar … it told me it was a carbonate … Because lemon juice is an acid and you can't do acid with acid … Sugar—it's a carbonate … because if it was an acid, it would have done nothing.' Some thought there was a middle category between an acid and a carbonate and drew on more than one observation: 'Flour's between acid and carbonates … I'm not sure … because when flour mixed with all liquids, it turned into a paste.'

After the second lesson, two of the six students identified that moisture was important for the reaction to occur between the acids and carbonates in sherbet: 'it was a mixture of the carbonate and the acids … they didn't fizz because there was no water … It only fizzed when you put them in your mouth.' Following the third lesson, students reiterated many of the points Patsy had stressed about observation, identifying some investigative as well as nature of science understanding: 'We had to touch and smell … And just like, keep your eyes on it'; and that it was important to look carefully 'Cos after, if you don't, you miss out everything … And you need to be patient … Because a reaction might be five or 10 min after.' They thought that scientists have to look really carefully for the same reasons. Students agreed at the end of the three lessons that the most important learning Patsy had wanted them to do was to learn to observe carefully 'that you have to be patient and observe every little thing that happens.'

The post-unit questionnaire was completed by 26 students after a two-week break following the last science session. Analysis of comments showed that ten students included that Patsy wanted them to learn about reactions or mixing things: 'how different things can mix and make another thing.' Seven indicated her intention was for them to learn about science and how science works. Fourteen comments included that air was inside the bubbles created in reactions, ten that it was a chemical and seven that it was a gas or carbon dioxide. Seven students thought that the bubbles were formed by a chemical reaction and another seven said the bubbles came from mixing baking soda and vinegar; three said they were caused by pressure, and two each by water, acid and dishwashing liquid. Most students said they observed (13) or mixed things (12) when doing science. Eleven said they do these things to learn, six to find out things and four 'so you know what happens'. What they associated with behaving like a scientist again reflected the points Patsy emphasised: observing (11); being patient (8); recording/taking notes (8); focusing carefully (5), although other unrelated factors also featured quite highly: being safe (6) and listening (5).

In summary, despite little discussion and clarification following investigations, some students developed useful substantive understanding about reactions, although many had confused or incorrect conceptions about acids and carbonates. Students' interpretation of their learning from the set of investigations in Patsy's class largely reflected her intent focus on the investigative skill of observation, the rigour of which they associated with doing science.

3.7   Summary

For the two teachers in these case studies, there was a strong congruence between beliefs about the nature and purpose of school science investigation and what happened in their classrooms. These teachers appeared to use science investigations to provide students with experiences in line with their beliefs about the nature of science, in particular, that it is evidence-based and relies on accurate observation. Less emphasis was placed on connecting students' observations with substantive ideas. Investigations were often teacher-selected, although in Alison's class there was a strong focus on student-owned investigations that was less common in Patsy's class. Investigation approaches that were included were mostly exploration, but with some experiences that involved classifying and identifying, exploration of a model, and one that involved investigating the effect of changing one variable—the beginnings of a fair test.

Both teachers provided students with learning opportunities that addressed the three major educative purposes for school science investigation identified in Chap. 1 through the science investigations they included. More attention was paid to developing investigative skills, and connecting student experiences explicitly to the nature of science than was given to developing substantive ideas. Key strategies observed in the teachers' practice are summarised below:

1. Developing investigative skills

   (a) providing enough specific equipment to enable all students to investigate in a focused way,
   (b) allowing time for exploration,
   (c) modelling wondering and supporting students to turn wonderings into investigable questions,
   (d) using questioning and feedback/feedforward to help students make investigations more orderly and rigorous,
   (e) restricting the range of variables to explore,
   (f) encouraging students to observe carefully.

2. Developing substantive understanding

   (a) providing specific equipment related to a specific concept,
   (b) focusing students on key observations,
   (c) connecting science ideas with observations,
   (d) using a model to demonstrate a science idea,
   (e) shaping class discourse to focus on accepted science ideas relevant to the investigation.

3. Developing an understanding of the nature of science and developing Science Capabilities for Citizenship

   (a) explicitly connecting students' scientific behaviours with those of scientists,
   (b) explicitly describing the way that scientists behave,
   (c) developing students' capability to gather data by supporting them to make detailed and careful observations.

Student learning in the Year 5/6 class reflected the teacher's emphasis on careful observation being important in science; some accurate substantive understanding of chemical reactions was developed, but some were confused about chemical properties of acids and carbonates. For younger students, the important learning from the unit appeared to be focused more strongly on what they had accomplished for themselves during the lesson in their own play/exploration. With strong teacher support, these young students' wonderings were turned into investigable questions that they could answer collectively, again with strong teacher support, using evidence from teacher-guided investigations and observations. With regard to learning about the nature of science, younger children connected their learning to the topic of bubbles, rather than to science. The older children, however, identified careful observation as being an important aspect of science.

Consider these

1. Why would you include science investigation in your teaching? What would you expect students to learn from it?
2. Consider a specific practical science investigation. What substantive learning could students develop from it? What investigative skills could they develop? What could they learn about science itself and how to engage with it?
3. Which strategies practiced by the teachers in this chapter would be useful to support the purposes you identified in Question 2?

Appendix 3.1: Lesson Analysis Summary of Alison's Bubble Unit with a New Entrant Class

		Lesson 1	Lesson 2	Lesson 3
Stated purpose			To answer individual children's questions about bubbles (developed from wonderings in lesson 1)	As for yesterday— answer children's questions
Teacher emphasis		Observation, what is left when the bubble bursts, experience of bubbles	As above	As above
Classroom environment		Pleasant, bright cheerful with lots children's thinking and work visible including bubble photos made into big book story on display.	As for lesson 1. Today children very tired after jump jam all morning	Less tired than yesterday

(continued)

(continued)

		Lesson 1	Lesson 2	Lesson 3
		Children on mat —reluctant to come to mat, sleepy after lunch—very new entrants. Outside for bubbles		
Lesson overview	Advance organiser	Programme for the day on the board, Teacher goes through where we are in the day	Programme run through as in lesson 1	Programme run through—Note: science is called Bubbles
	Shared learning intention		Let's see if we can find out about these questions	Shares question to be investigated and activity
	Made link to previous lesson	Started by re-reading the bubble story and postcard from the postcards resource. Read out their bubble ideas from previous lessons	Strong links to previous lesson; children's questions/ wonderings	Links to yesterday
	Checked home-work		Listens to links child has made at home— Bobbie answered his question!	
Contextualisation Evidence of connecting with students; students world; culture		Story and postcard. Previously children have read bubble stories and completed 'what we know about bubbles chart' Some exploration using bubble gun, completed class photo story	Student questions guide the lesson	Used Bobbie's balloon he brought for news—like bubble because you can fill it with air. Teacher knew about the balloon and used it in the lesson

(continued)

(continued)

		Lesson 1	Lesson 2	Lesson 3
Evidence of relationship		Teacher knowledgeable about each child's needs and strengths	Each student's question written on a card. Listens to what they have done at home— children share ideas with teacher	

Investigation

	Lesson 1	Lesson 2	Lesson 3
Type	Free exploration	Exploration	Exploration
How was it introduced?	Story used to determine focus 'I wonder if we will see anything left on the ground?' then, student-driven: bubble mixture put out one per child and they follow their own curiosity blowing bubbles. Going to 'be scientists'	Student questions on cards	Reading from a scientific text about bubbles and how they burst
Skills Taught/ practiced/ applied [be specific]	Observing carefully: taught through imagining inside before going outside, then opportunities for practice outside	Focused observation-practiced. Practiced blowing bubbles without mixture first	Focused observation; colours in bubbles— outside and on torch in the dark: what colours can you see?
Capability Taught/ practiced/ applied [be specific]	Observation-practiced Teacher asked them to focus on: what's left after the bubble bursts —is there anything on the ground like in the book? What do you think has happened to the bubbles?' What do you think is I the puddles? What do you think would be left	Gathering/using evidence to answer their questions-with strong support and guidance!! Student question: can you catch a bubble on your finger? Teacher: who caught a bubble on their finger? Student Question: Can bubbles be different shapes?	As above: observation but with links to vocab; 'I am looking for lots of colour words!' Reinforced with lots of praise 'I'm so proud of you and your looking today!'

(continued)

(continued)

	Lesson 1	Lesson 2	Lesson 3
	over from the bubble mixture? Supported with: 'Good explanation!'; 'Good talking'; 'What a good scientist you are!' Magnifiers offered but not much interest— too busy with bubbles! Also deliberate scaffolding/ pushing for development of questions from wonderings Lots of vocabulary development from science	Teacher had provided different shapes to blow bubbles with— they identified different sizes but did not get different shapes	
Teacher-led/ demonstration	No—apart from choice of bubbles topic and leading imagining of observation of bubbles popping	Teacher led to answer children's questions	Teacher led from children's questions but also her own focus on bursting
Instructions Verbal/ written/text	Verbal	Verbal	Verbal
Student involvement Decision about what to do:	All students were allowed to play with the bubble mixture: organisation allowed them each to have their own mixture and they each explored independently. Teacher suggested some things to look carefully at—focus was on what happens when the bubble bursts. 'Look carefully at the ground where it bursts.'	Student questions— answered with strong guidance from teacher but students involved as much as possible— e.g., Aggie gives out equipment as it is her question to be investigated Some students (advanced readers) have made a different mixture from school journal which is also investigated	Student questions but also teacher idea for activity

(continued)

(continued)

	Lesson 1	Lesson 2	Lesson 3
Investigate		Yes	
Choose equipment	No	No	No
What to measure	NA	Invited to suggest how they could find out e.g. why do they pop? Teacher: How could we find out? Child: We could look at bubbles... Child 2: Could look in a book... They do the next day! Books as referents?	No
How to record results	No	No—simple observation	No
How to analyse/ summarise	No		No
Time given to think	Time to play and explore	Time to play and explore	Time to play and explore
Critique their own design			
Critique others design			
How was the investigation concluded Reflection:	Reflection— considerable time spent by teacher recording individual 'I wonder' statements. Children got a bit bored and restless on the mat while other did theirs	Reflection—using evidence from observations to answer each child's questions	Reflection—children draw bubbles and the colours they saw
Quality of engagement	Enthusiastic: all!! Perseverance: most Attentive: most	High during practical outside, but lost interest in any talk time after about 5 min —rolling round on carpet During investigation: Enthusiastic: most Perseverance: most Attentive: most	Enthusiastic: most Perseverance: most Attentive: most

(continued)

(continued)

	Lesson 1	Lesson 2	Lesson 3
On-task behaviour	Most	Most during investigation some during talk time afterwards	Most
Researcher's view	Focus: observation and fostering curiosity Students learnt 'What happened when I…' for bubbles. They followed their own lines of interest and observation, some with long intense periods of concentration: trying to join two bubbles together or trying to catch a bubble on a leaf or stick Children making useful observations in response to teacher Question: 'When they pop they never come back' 'When some of the bubbles popped something was left'	Focus was on gathering evidence to answer their own questions Children were interested in their question but not others… child on tape was interested in HER question and what she did to find out Children care and know whether or not their question has been answered- one child interviewed had not had her question answered—thought they would next time…	Children were able to give specific detailed observations about the colours they had seen 'I can see a rainbow!' 'I can see purple!!'

Appendix 3.2 Student Post-unit Questionnaire—Patsy's Fizzing and Foaming (Acid/Carbonate Reactions Unit with Year 5–6 (8–10 Years Old) Students

Question 1   What do you think your teacher wanted you to learn from doing the science activities?

Question 2   You are telling a new classmate how to behave like a scientist when you are doing an experiment in class. What are three things you would say?

Question 3   When we mixed up things together we saw some bubbles or froth.
   a.   What do you think was inside the bubbles?

   b.   What do you think made the bubbles happen?

Question 4   What sorts of things do you do when you do science?

Question 5   Why do you think you do these things?

Question 6   If we did not do experiments in science what would we learn?

Question 7   How do you know what you are doing is science?

Question 8   What would you not do in science?

References

Abrahams, I., & Millar, R. (2008). Does practical work really work? A study of the effectiveness of practical work as a teaching and learning method in school science. *International Journal of Science Education, 30*(14), 1945–1969.

Anderson, D. (2013). Leading change in primary science: Experiences of Primary Science Teacher Fellows who have raised the profile of science in their schools. *Journal of Educational Leadership, Policy and Practice, 28*(2), 15–27.

Anderson, D. (2015). The nature and influence of teacher beliefs and knowledge on the science teaching practice of three generalist New Zealand primary teachers. *Research in Science Education, 45*(3), 395–423.

Appleton, K. (2006). Science pedagogical content knowledge and elementary school teachers. In K. Appleton (Ed.), *Elementary science teacher education: International perspectives on contemporary issues and practice* (pp. 31–54). Mahwah, NJ: Association for Science Teachers & Laurence Erlbaum.

Bull, A. (2015). *Capabilities for Living and Lifelong Learning: What's Science Got to Do with It?* Retrieved from http://www.nzcer.org.nz/system/files/Capabilities%20for%20living%20and%20lifelong%20learning%28v2%29.pdf.

Bull, A., Gilbert, J., Barwick, R., Hipkins, R., & Baker, R. (2010). *Inspired by science* (A paper commissioned by the Royal Society of New Zealand and the Prime Minister's Chief Science Advisor). Retrieved from http://www.nzcer.org.nz/pdfs/inspired-by-science.pdf.

Chamberlain, M., & Caygill, R. (2012). *Key findings from New Zealand's participation in the Progress in International Reading Literacy Study (PIRLS) and Trends in International Mathematics and Science Study (TIMSS) in 2010/11.* Retrieved from http://www. educationcounts.govt.nz/__data/assets/pdf_file/0011/114995/Key-Findings-NZ-Participation-in-TIMSS-and-PIRLS-2010–2011.pdf.

Education Review Office. (2004). *The quality of teaching in years 4 and 8: Science.* Retrieved from http://www.ero.govt.nz/National-Reports/Science-in-Years-5-to-8-Capable-and-Competent-Teaching-May-2010/Introduction/The-quality-of-teaching-in-Years-4-and-8-Science-2004.

Education Review Office. (2010). *Science in Years 5 to 8: Capable and competent teaching.* Retrieved from http://www.ero.govt.nz/National-Reports/Science-in-Years-5-to-8-Capable-and-Competent-Teaching-May-2010/Overview.

Education Review Office. (2012). *Science in the New Zealand Curriculum: Years 5 to 8.* Retrieved from http://www.ero.govt.nz/National-Reports/Science-in-The-New-Zealand-Curriculum-Years-5-to8-May-2012/.

Fitzgerald, A., Dawson, V., & Hackling, M. (2013). Examining beliefs and practices of four effective Australian primary science teachers. *Research in Science Education, 43,* 981–1003.

Hodson, D. (2014). Learning science, learning about science, doing science: Different goals demand different learning methods. *International Journal of Science Education, 35*(15), 2534–2553.

Millar, R. (2010). *Analysing practical science activities to assess and improve their effectiveness.* Hatfield: Association for Science Education. Retrieved from http://www.york.ac.uk/media/educationalstudies/documents/research/Analysing%20practical%20activities.pdf.

Millar, R. (2011). Reviewing the national curriculum for science: Opportunities and challenges. *Curriculum Journal, 22*(2), 167–185.

Ministry of Education. (2001). *Making better sense of the material world.* Wellington, NZ: Learning Media.

Ministry of Education. (2007). *The New Zealand curriculum.* Wellington, NZ: Learning Media.

Mortimer, E. F., & Scott, P. H. (2003). *Meaning making in secondary science classrooms.* Maidenhead: Open University Press.

Paul, R. (2007). *The king's bubbles.* Auckland: Scholastic.

Shulman, L. S. (1986). Those who understand: Knowledge growth in teaching. *Educational Researcher, 15*(2), 4–14.

Treagust, D. F. (2007). General instructional methods and strategies. In S. K. Abell & N. G. Lederman (Eds.), *Handbook of research on science education* (pp. 373–391). Mahweh, NJ: Lawrence Erlbaum.

Chapter 4
Science Investigation in Secondary School

New Zealand aspires to be a country where science and technology play a vital role in its economic growth and the well-being of its citizens, and science education is an integral part of this vision. The educational curriculum requires students to engage with, explore and experience the natural and physical world with the aim of enabling them to 'become critical, informed and responsible citizens' (Ministry of Education, 2007, p. 17). Teachers are set the task for this to take place in their science classes. Research suggests that curricula are aspirational and focus on the ideal and that the taught and learned curriculum are not always the same as the mandated curriculum (Fernandez, Ritchie, & Barker, 2008; Hume & Coll, 2010).

Teacher beliefs are reported as a major influence on the development of knowledge for teaching (Anderson, 2015). Research suggests that teacher beliefs are a strong driver in the implementation of the curriculum and, therefore. that the science students experience and learn is guided by the teachers. Therefore, teacher beliefs are considered an important consideration in all teacher professional development programmes (Lowe & Appleton, 2015).

Here, we draw upon case study research investigating the teaching and learning of science investigation in a secondary school with a focus on the complexity of teaching and learning science investigation to gain deeper insights into what teachers believe science investigation to be, how they teach it and what students learn from investigating. The research was conducted at Central School (pseudonym), a large coeducational school located in an urban area. The school had approximately 1100 students. The participants comprised three teachers and their year 9 (age 13 years, which is first year of high school) classes. Central School has a strong science department where teachers have opportunities for professional development and are aware of the recent policy changes and online resources provided by the Ministry of Education.

The topic during the time of this research was *Why water?* The intention was for students to learn about the particle nature of matter and change of state, including the water cycle, solubility, physical and chemical change. As this was a Year 9 class, it was intended that students would become familiar with basic laboratory

© Springer Nature Singapore Pte Ltd. 2018   71
A. Moeed and D. Anderson, *Learning through School Science Investigation*,
https://doi.org/10.1007/978-981-13-1616-6_4

equipment, and learn how to use it. The plan included achievement objectives from the Nature of Science strand of the curriculum at this level which was reflected in the unit plan that had a focus on science investigation. The teachers wanted to introduce the recently developed Science Capabilities for Citizenship designed to help teachers to more purposefully integrate the nature of science in their teaching (Refer back to Chap. 1), particularly, focussing on making observations, collecting data and making evidence-based explanations.

There are three aspects to this chapter, which are given as follows:

- Teacher beliefs about the nature of science investigation and why students should be taught to investigate,
- Teacher practice of teaching science investigation,
- Student learning through engagement in science investigation.

We start the chapter with the beliefs held by the three teachers, Hazel, Chris, and Fran (all pseudonyms) at Central School, who participated in the research. All three teachers were experienced teachers of science.

4.1   Teacher Beliefs About Nature of Science Investigation

Hazel held a strong belief that students needed to know why they do an investigation and that ideally, it has to be student led. She highlighted the need for students to be actively involved, and have '*fiddle time*'. In her classes, she wanted to develop a culture of learning together and from each other, and generate ideas to investigate. She said that students needed to develop the skills to plan an investigation, control variables, measure accurately, and to come to evidence-based conclusions. In her view, the teacher's challenge is selecting something that the students could do in the classroom.

Chris believed that science investigation should be at the heart of science education. He said that everything that we do and everything that we think should be predicated based on real-life actual investigations in science:

> It involves hands-on work by the students. It involves teacher, often, the teacher, me, knows what it is and the direction and the concepts that I'm trying to get across. It is actually directed but then as much as possible and increasingly … I will get them to select the equipment from the range available to work out their aim and their method and, and they do it, rather than I tell them in a step by step recipe labelled sort of way…. work out an experiment which actually tests what you want to find out.

For students, the most important thing is to find what *they* want to find out and do what *they* want to do, not what the teacher thinks. Chris wanted students to investigate 'what they want to know about, what burns them up, it is coming from their angle.'

Fran said that she gets her students to do either research-type investigations or practical investigations because, in her view, there could be three approaches to science investigation:

> For me there can be three approaches to investigation: get the students to find out information through inquiry, so it would be research based; I could get the students to do a practical investigation to find out for themselves something about a scientific/biological concept or process; or it could be finding out about the properties of something in which case I would give them some material and either give them a process to follow, or for them to work out how they were going to do the practical to answer the question.

All three teachers talked about their role in students' investigation and they agreed that it was the teacher's role to present the problem and to look for opportunities to engage students in science. Hazel said that ideally, what students investigate ought to be selected by the students and pragmatically, her practice was to model an aspect of the investigation for the students to decide what they wanted to investigate, for example, she would model the effect of the surface area on the rate of reaction of enzymes, and then students could investigate the effect of temperature or concentration. She added, whatever the teacher plans for the students to investigate in Years 9 and 10 needs to be simple, it can be completed in an hour, and it is accessible to all students.

Fran talked about a recent investigation she did with her science class. She began the topic on genetics with taking the 'magic school bus' through the body and starting from the skin, 'drilling down to cells', then into what might be found in the cell, the nucleus, what might be there, down to chromosomes and genes and DNA. She wanted them to find out for themselves what the skin cell might look like, and they extracted the DNA from their own cheek cells. She says that she understands the concepts associated with cells are abstract and the students do not actually have any idea of the size of these things and so the first thing is for the students to make:

> their own slides, learn to use the microscope ... When I was their age, all I could see were my own eyelashes... It is important that they learn to use the microscope because it provides access to the world on a totally different level.

During teacher group discussion, Hazel explained the purpose of using demonstrations and her ideas about the nature of science investigation:

> A demonstration implies that there was something, there is some information that I want to get across to students and I'm going to show you this in my demonstration and visual is often the way that, you know, visually students respond to that. An investigation implies that they're finding out something and they don't know.... The students are on a, a voyage of discovery. They're investigating. The difficulty we have as science teachers, of course, is that we have to keep the parameters safe.

Chris sees his role as the facilitator who helps the students to learn what they need to learn about the subject he teaches and makes it possible for them to learn what they are passionate about, providing the resources for learning, materials to work with, technology to support and record evidence (for example, making audio and video recording devices accessible), and who challenges them to think and experiment. He gave the following example:

> At one time we had guys who were interested in motor mechanics so they were actually building Lego models of cars and trucks and trying to work out forces and we had a teacher aide who was actually running programmes on year 11 Physics for the kids. So these might be low-achieving boys but they are passionately interested in mechanics and motors and he is showing them mechanics. They don't know it but they were actually doing a Year 11 mechanics course.

Fran was passionate about her students becoming scientifically literate. She believed that school science can be like scientists' science, depending on what science is done with the students. For example, her mother is an ecologist who investigates water quality and what animal, or plant life may be present in a stream that would indicate the health of the stream. Fran and other teachers in this school take the students to the local stream and investigate the quality of water there. Fran gave the following example to explain how she encourages her students to be critical like scientists:

> I think we can challenge our students to think about aspects of investigation that scientists consider, for example, if we surveyed mothers in the Waikato (a region in New Zealand) to find out the level of vaccination in New Zealand, can we do that? Why might it not be a representative sample? I challenge them to think about what is being claimed, who is making the claim? What evidence are they using to make these claims?

Even if school science may not be *exactly* the same as the science scientists do, students can be given the opportunity to learn how real scientists *do* their science. Fran aspires to give these opportunities to her students. In her view, not all of her students will go out there and become scientists but they do need an understanding of the science in their everyday lives, including how science is done.

All three teachers held a strong belief that students should have the opportunity to investigate. They demonstrated a nuanced understanding of the empirical purpose of science investigation and wanted their students to understand that science investigations do not always progress in a linear fashion and that investigating is a messy and iterative process. Chris, for example, wanted his students to understand that 'science is a progression of failures, magnificent, spectacular, disastrous failures', and that investigations do not always proceed as planned. The teachers all explained that science investigation does not always have to be practical, it could involve researching for information. They wanted their students to realise that school science investigation is an important approach to learning science. From a practice perspective, they selected a science investigation depending on what they intended the students to learn: learn/master a skill, understand a science idea, and/or understand how scientific knowledge is created.

4.2 Teacher Practices that Support Student Investigation

To gain a deeper insight into teacher practice, three lessons were observed in each teacher's class. As the focus of the research was science investigation, all observed lessons involved aspects of students carrying out investigations. Although all three teachers had very different teaching styles, all lessons observed had the following similarities:

- Each lesson had a clear focus, with teachers sharing the learning intentions with the class and making links with the previous lesson.
- All lessons provided students with the opportunity to engage in science investigation.
- Each lesson ended with a reflection phase where students had the opportunity to talk about their investigation, followed by teacher-led discussion.

Investigation in Hazel's Class: Separating Mixtures

Hazel's class started the lesson with a quick 5–10 questions to make connections with the previous lesson. Learning intentions were shared with the students verbally and her practice was to have the key science vocabulary listed on the board. The first observed lesson was teacher-led and students were given instructions on what equipment to use, what they needed to record and how they needed to record it. Students were engaged in dissolving a range of household products in water. The purpose was to give them practice using equipment, make observations, and offer explanations. This was followed by their mixing two liquids, measuring accurately, and making observations. At the end of the lesson, the results were discussed, and students completed the lesson by identifying the solute and solvent in each case and defining solubility. This was related back to the model of solid, liquid and gas that they had learnt in a previous lesson.

In the second lesson, students learnt about diffusion in liquids and gases. The lesson required students to observe diffusion of potassium permanganate in water, which was followed by two demonstrations of diffusion in gases. As with lesson one, the second lesson had high student engagement. All students were able to complete the task and again the lesson ended with discussion, this time about particle nature of matter and how diffusion works in liquids and solids. The lesson observations were made early in the year and the teacher wanted the students to learn to use the equipment and develop vocabulary, and the focus was on the first Science Capability for Citizenship, gathering and interpreting data, where students are required to make careful observations and learn to differentiate between observation and inference. For example, the following statement highlights the difference between observation and inference:

> Science knowledge is based on data derived from direct, or indirect, observations of the natural physical world and often includes measuring something. An inference is a conclusion you draw from observations – the meaning you make from observations. Understanding the difference is an important step towards being scientifically literate.[1]

This was the third observed lesson. Having learnt the practical skills of separating solids, the objective of this lesson was for students to plan and carry out an

---

[1]http://scienceonline.tki.org.nz/Science-capabilities-for-citizenship/Introducing-five-science-capabilities.

investigation to separate a mixture of salt, beans, rice, sand and iron sand. Evidence suggested that the students had learnt what mixtures were and could apply the various procedures they had learnt to separate them.

After planning their investigation, each group decided on the equipment they needed. As they separated each component of the mixture they had to put it in a watch glass and bring it up to the front of the room where the teacher had drawn up a table on the table top. As each group progressed, Hazel was able to keep an eye on the table and see which groups were lagging behind, going around the class to talk and check with the students and offering them support as required. There was a bit of a competition going—how long did each group take to separate their mixture?

Hazel   Keeping an eye on the table. Group 6 "where is your salt?"
Student   We have too much water and it will take a long time to evaporate it. Can we
          leave it in the evaporating dish?

Before the end of the lesson, all groups completed the task and tidied up. The focus of the lesson was to plan and carry out an investigation and the intended learning was application of the procedural skills the students had learnt in the previous lessons. They were also applying their knowledge about the solubility of solids in water that they had been learning. There was a quick recap in which the students explained the following:

Use magnet to separate the iron sand

↓

Pick out beans with tweezers

↓

Sieve the sand and separate the rice

↓

Add water to the sand and salt to dissolve the salt

↓

Filter out the sand

Evaporate the water to recover the salt.

Homework was to watch a Ted talk (a video) on their netbooks about water purification (8 min).

As the three observations reported earlier were not made in consecutive lessons, the researcher did not see how the separation techniques were taught and what input the students had in learning those skills. In each lesson, Hazel created an opportunity for students to see how what they were doing in class was *science in their everyday life*.

An Investigation in Chris's Class: Exploring Ice

Chris believes in giving students the opportunity to explore and find answers to their own questions. His style was to create learning possibilities and allow students to follow their interest within the constraints of the intended learning for the topic. Chris was well-organised for each lesson with learning intentions on the whiteboard and greeting and inviting students in. The classroom was set up with a lei (garland) on some desks and sparkling hats on others. Resources were prepared and put out on the side benches. The atmosphere appeared 'festive', unusual, and welcoming. The students, when they arrived, were excited and settled down quickly, waiting in eager anticipation (observation notes, lesson one). Chris engaged with students as they came in, a bubbly, enthusiastic Year 9 class of 20 boys and girls.

In the first lesson, students investigated ice; in lesson two, they explored dry ice; and in the third lesson, they looked at neutralisation and dissolving. Lesson one began with three students putting on hats and going up to see the teacher who handed each one of them a video camera. When asked what they were going to do, the students said they were allowed to gather digital 'evidence' when they did experiments. If they wanted to film other students' experiments, they went to the group who could see the person with the hat coming towards them. If they did not want to be recorded they just put their hand up and then their choice was respected and they were not filmed. Later in the lesson, the recording was uploaded to a secure website that all students had the opportunity to view. Other students could provide feedback through blogs. Chris said he made sure that students with the 'sparkly hats' returned the hats and the cameras before they left the class. The strategy appeared to work well in this class. Similarly, the colourful lei were worn by the student in each group who collected and returned the equipment. Over the three lessons, it was clear that students made sure that the garland and the cleaning up were equitably shared.

Chris said, 'I want you to sit quietly and think about everything you know about ice. You have 2 min to do this.' Students responded by quietly writing down what they knew about ice. This was followed up with a class gathering of information on the whiteboard. Next, the teacher said, 'Now I want you to think about something you not know about ice and would like to find out.' They were given a couple of minutes to do this. Students then had to phrase their thoughts as questions they could investigate. These questions were collated on the board and the students who wanted to investigate similar things got together in groups of three. Chris already had the resources students were likely to need on the side benches. One group wanted to investigate if there was a difference in the way ice 'shattered' when hit with a hammer. Chris found them something heavy from the preparation room to use as a hammer. They could also choose small ice cubes or larger blocks of ice. Overall, this lesson was a student-led investigation where the context was set by the teacher. For the next 20 min, eight very enthusiastic groups of students carried out their investigation. A summary of their activity is presented in Table 4.1.

Analysis of lesson one is presented in detail as this was a lesson that illustrates how groups of students could identify their own questions and plan an investigation, carry it out, process their data and share their findings within one lesson. The summary is from the analysis of the classroom observations which were audio-recorded and transcribed. The last column identifies relevant science capabilities for citizenship.

Ten minutes before the end of the lesson, students stopped and tidied up. They digitally uploaded their investigation and pictures onto the common drive for others to visit. Each group shared their findings with the class. This investigation afforded students the opportunity to consider what they already knew, and they selected the investigation they wished to carry out. The reflection at the end by the students, although brief, was guided by the following framework provided by the teacher:

What did you want to find out?

How did you do this?

What did you measure/observe?

What did you find?

For homework, Chris asked students to check an investigation uploaded onto the shared drive that they wanted to find out more about and ask any questions that arose through a blog.

In lesson two, students explored dry ice. As they had been studying water, the purpose of investigating dry ice was at first puzzling. However, the teacher made the link between the lessons clear because he discussed the change of state of water in lesson one where students had seen it change from ice, to water, to steam. The use of dry ice was his way of bringing in the change in state from solid to gas and introducing the idea of sublimation. The lesson involved an exploration of dry ice guided by the activities that were set up around the room. Students had a worksheet that explained what they were to do and were asked to write their observations on the sheet. As with lesson one, some chose to digitally record their findings.

The focus was to do the activity, make observations, and write inferences. The teacher was very concerned about safety [rightly so!] so there was less focus on learning. In one task students were adding $NaOH$ to the water which had dry ice in it. The intention was for students to observe the process of neutralisation. Through moving around the class, Chris worked out that students did not 'get' neutralisation and that this would be followed up in the following lesson. He commented that the learning from dry ice exploration was limited because students did not have previous experience with dry ice. During the interview, he said that because students did not have any prior knowledge about dry ice, they needed two lessons, one to explore dry ice and the following one to decide on what they wanted to find out about it through investigation. Chris added, 'Dry ice is expensive, we usually get it on one day.' Then, he added, 'it would be better to do dry ice on the day we have a double period.' The third lesson was different from the one he had planned because Chris wanted students to have more time investigating neutralisation to sort out the intended learning.

Table 4.1 Detailed analysis of lesson one, investigating ice

Investigation type	What students did	Processing information	Looking at evidence	Findings	Science capability
Exploration Find out how long it takes for an ice cube and a large ice block to melt	Heated a small ice cube and a large one to find out which takes longer	Made observation and inferred that not all ice had to melt for the melted water at the bottom of the saucepan to boil	Did not repeat investigation but knew that to be sure they would need to. Pointed to the group who were also doing the same investigation and had similar results	Were surprised that some of the water started to change into steam while there was still ice in the saucepan	**Gather and interpret data** **Use evidence** (measured time taken for the blocks to melt)
Exploration Find out the temperature at which ice melts and water boils	Heated ice cubes and measured temperature	Each member of the group read the thermometer and checked the temperature when all the ice had melted	Stopped heating when taking and checking the temperature of melted water. Started heating after measurement	Students were doing this next to the groups above and were surprised that they could see water turning to vapour before the water started to boil	**Use evidence** (measured the temperature at which all the ice had melted and the time taken for the blocks to melt)
Exploration Find out if vapour and steam are gas or water (Two groups)	Boiled water in a jug and observed the vapour escaping	Hand got wet. Steam was invisible	All four had felt the water vapour. Agreed that steam was invisible	Decided vapour was water and steam was gas	**Gather and interpret data** **Observation and inference**
Pattern seeking Find out how small and large ice cubes break when dropped from different heights	Dropped a small cube from level one of the building, took a photo of how it shattered, repeated with a large margarine container size cube. Took photo	Back in class, looked at the photos for both sized blocks and compared patterns. Each student looked for patterns and agreed that the size of the block did not seem to make much difference in relation to the	Came up with a new question: Wonder how the large block would break if it was dropped on its side? Wanted to investigate this in the future	Found that the small cube shatters into tiny bits but the large one broke into one large piece and a lot of smaller pieces	**Gather and interpret data** **Observation and inference** **Critiquing evidence** Making generalisations based on multiple trials

(continued)

Table 4.1 (continued)

Investigation type	What students did	Processing information	Looking at evidence	Findings	Science capability
	Repeated from levels 2, 3, 4, 5	height it was dropped from			
Pattern seeking Finding out how ice shatters when you hit it with a hammer	Large block and small cube did this outside in the corridor, took photos	Lot of fun but realised that they did not have the evidence to come up with a conclusion		Decided there was not really a pattern, depended on who was hitting it and how hard they were hitting	**Critiquing evidence**
Making a model Making an igloo	Sprinkled salt on ice cubes. Pushed them together and held them. One boy decided to put salt and ice in his hand and closed his fist, claimed he had burnt himself		In the focus group: Wonder how long it took the Eskimos to think about building a house without cement	Said salt melted the ice and pressing together froze the water. Made an igloo! Took a photo	**Gathering and interpreting data Observation and inference**

An Investigation in Fran's Class

As with Hazel and Chris, three observations were made in Fran's class. Fran arrived in class early and set up her classroom with learning intentions on the board and a list of *must do* tasks and *could do* tasks. The first lesson was for students to plan and carry out a fair testing type of investigation (discussed in detail later). In the second lesson, students investigated the effect of temperature on ice. Their task was to draw a table and record change in temperature and later to graph the results. The last 10 minutes of the second lesson was used for students to make presentations. In the third lesson, students investigated evaporation and condensation by heating some water in an aluminium can until it was full of steam and then turned it upside down into a container which resulted in an imploding can. The focus here was to make observations and explain the observed phenomenon.

Although all three investigations were teacher-led, in lesson one students had to plan their investigation and decide on the equipment; in lesson two, they were given instructions about how to set up the equipment but were asked to create an appropriate table to record their results. In lesson three, they first observed a teacher demonstration and for the second activity, Fran showed students what she wanted them to do. The activity followed a Predict Explain Observe Explain format. They had to predict what observations they might make as the water boiled in the can and when they would turn it upside down in cold water. Students were asked to provide explanations for their predictions.

This was the first investigation and was posed as a problem. The context was the various soft drinks the students drank. They were to find out how much sugar each of the drinks contained, and this had to be carried out so that it was a fair test.

Each group was given a can of soft drink (coke, diet coke, lemonade, diet lemonade and Fresh up fruit juice). Students were given time to work in groups and plan their investigation and decide on the equipment they would use. They could use 50 ml of the drink they were investigating. Each group would weigh an empty evaporating dish, put 50 ml of drink in it and heat it to evaporate the water. They set aside the evaporating dish to cool and to weigh in the following lesson. The results of each group were collated on the board. Fran focussed on teaching students the science capability of *critiquing evidence* through this investigation. This was done through a series of questions posed to the whole class followed by students being asked to look at the cans of drink that their group investigated. Fran talked the class through the information on the can.

Fran   Compare your results with those on the can. The information on the can is per 100 ml, you used 50 ml. Work out how much sugar would be in 50 ml. Consider why there may be a difference between your results and those on the can.

S1   They would have checked the amount of sugar in a large number of cans and averaged.

She emphasised on how multiple trials add to the robustness of the investigation design and make results more trustworthy and asked them to do the following:

Fran   Write down how you can have more confidence in your results. Then as you finish this do the *can do* task

        1. Work out how much sugar there may be in the can of drink.
        2. List the drinks from most sugar to least sugar.

The reflection was done as a teacher-led discussion guided by the following questions:

1. Did the liquid look different to how it looked before? Why might that be? [Observation/inference]
2. Did anyone get any crystals?
3. How was this solution different to the copper sulphate solution? [Thinking scientifically] [Saturated; drinks not saturated drinks]
4. Who had the result close to their prediction?
5. How did your results differ?
6. Which would be bigger, a pile of sugar or Nutra sweet? [Made links with diet coke/coke from the previous lesson].

Fran also talked about phenylalanine [labelled on the drink container] and its impact on metabolism where her focus on biological literacy came to the fore. Then, she moved around the class and checked students writing up: Aim, hypothesis, observation, conclusion. It was a well-organised lesson and ended with refocusing on what else would the students like to investigate? May be other drinks?

The focus was on evidence and evidence-based conclusions. Students were thoughtful and analytical. The point not raised by either students or teacher was the possibility of there being something else other than sugar in the liquid they evaporated to be certain that what they were left with was sugar and not any other solid.

Lesson two had several learning intentions. Fran told the class that they would find out the temperature at which ice turns to water and water boils. This is a Year 9 class in the first term, so the teacher gave them practice to set up equipment, measure temperature, record temperature change in a table and write down their conclusion. She took the opportunity to first find out what students knew about change of state by getting them to draw and complete a diagram (see Fig. 4.1).

In the last 10 minutes of the lesson, students presented their homework task, which was to communicate eight to ten facts about water to the rest of their class. The teacher praised them for the creative ways in which they communicated the science ideas to their class (see Fig. 4.2).

The purpose of this presentation was to encourage students to find eight to ten facts about water and then to communicate these in a creative way. The intention was for them to learn that *scientists also share information and that it can be presented in a variety of ways.*

Overall, in Fran's practice, there was a strong focus on the following:

- learning science ideas,
- making observations and inferences,
- gathering and processing data,
- making evidence-based conclusions,
- critiquing evidence.

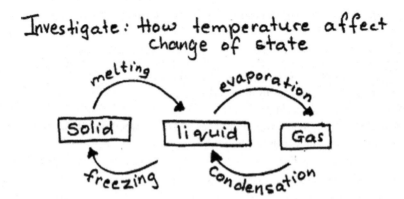

Fig. 4.1 Students completed the drawing by writing the change of state in each case

Many ways to communicate in science......
Powerpoints

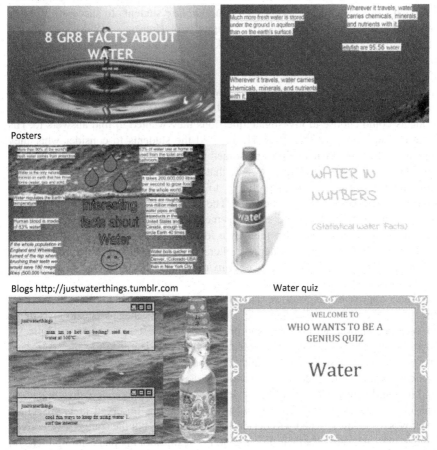

Posters

Blogs http://justwaterthings.tumblr.com   Water quiz

Fig. 4.2 Examples of presentations made by Year 9

4.3   Student Learning from Investigation

Supporting Student Investigations
As described above, all three teachers provided opportunities for students to investigate. Although we asked to observe lessons in which students carried out investigations, the analysis of documentation and student work showed that practical work was a frequent approach in these classes. As all three teachers were from one department in the school, there was evidence of collaborative planning of the units of work. However, even though all teachers engaged their students in investigation they had different approaches. For example, Hazel believed in students first developing the skills, and then being given the task to plan and carry out

their investigations. It was evident from the questioning in her classroom that students learnt the science ideas she wanted them to learn and developed the intended skills. She had a strong focus on *enabling* students to successfully carry out their investigation. This approach aligns with Hodson's (1998) view that if students do not have the basic skills, a lack of skills can become a distraction during investigating.

Chris's approach was closely related to a guided-discovery approach. He encouraged the students to ask questions they were interested in investigating and very capably provided the materials they would need. He set the context and within that students could choose to investigate questions they wanted to find answers to. His approach was not *anything goes*. He focused the students to consider what they already knew and then to find out something that they were interested in. The reflection phase was crafted to ensure that all students listened to each other's investigations and had the opportunity to hear what and how the others had investigated, and what they found out. The follow-up of uploading their investigation to a private website and encouragement to view things empowered students to take some responsibility for their own learning. The environment was festive, engagement high and the teacher was there to listen, ask questions, provide the resources and ensure that there was *thinking behind the doing* as suggested by Abrahams (2011).

Fran's approach was to draw students' attention to a problem, be critical about the information presented to them, develop skills and ask the students to investigate to seek an answer to the problem presented by the teacher. For example, in investigating the sugar content of soft drinks, students learnt to use the scales and weigh the evaporating dish before and after; they also measured an accurate amount of liquid while the teacher moved around the class helping them, asking questions, and it would appear picking up the issues as they arose. These were then unpacked and clarified during the reflection. In all investigations observed in Fran's class, what was to be investigated was decided by the teacher, but the students were given time to plan and carry out their investigation. The questioning had a strong focus on developing the capabilities of gathering data, looking for evidence, and critiquing the evidence. Fran's strong belief in scientific literacy and teaching to ensure that her students learnt to make informed decisions about the information presented to them took primacy.

At the department level, Hazel, the Head of the Science Department, encouraged professional development and all three teachers had been involved in it. Investigations were used for developing science ideas, procedural knowledge and an understanding about the nature of science.

Multiple data sources were used to gather insights into student learning. These included:

- focus group interviews,
- end of topic assessment results,
- student questionnaire.

Focus Group Interviews

A group of six students from those who had agreed to participate in an interview at the end of each observed lesson comprised students with a range of abilities (selected by the teacher) and there were three males and three females in the group. All six students from each of the three classes participated in all three interviews related to their class. The interviews were transcribed and the research team collectively read and individually coded one common transcript. A discussion followed and codes were reviewed and some combined. The Science Capabilities for Citizenship framework was used for generating the codes for the nature of science ideas (refer back to Chap. 2).

The focus group data show that the students were learning science ideas, for example, in Hazel's class (Class 1), there were 51 responses that were related to demonstrating understanding of a substantive science idea. In Chris's class (Class 2), there were 31 content related responses, and in Fran's class (Class 3) there were 23. From the students in the focus group, when probed there was evidence that students were making the link between the domain of objects and observables and the domain of ideas which is one purpose of engaging students in practical work (Abrahams & Millar, 2008). Content knowledge was not a focus of this research, however, students demonstrated understanding of the content. The limitations were that we only know about the students from each class who participated in the focus group. Other limiting factors could be the direction of the discussion during interview and how articulate the focus group was, as well as the teacher's focus during the lesson (Fig. 4.3).

When the science capabilities framework was used, we found that in most cases there were more capability 1 related responses in all three classes. This was also reflected in the observation notes as the emphasis in all three classes had been on making observations and inferences (Capability 1). And as the teachers were focussing on the capabilities, students appear to be learning them and using them during the focus group interviews (see Fig. 4.4).

End of Topic Assessment Result

Students in all three classes completed an end-of-topic test. One question relating to the content was taught during the nine observed lessons, and a second question that closely related to an aspect of investigation was carried out in each class along with the grades allotted to the relevant questions. The first question was intended to assess students' understandings about the particle nature of matter and change of state. The second question was a scenario presented to the students which required them to critique a plan and comment on whether the evidence presented would lead to the conclusion drawn. Why? Why not? These provide some evidence of science ideas learnt and retained by students over a few weeks. There are four possible grades that align with grades awarded in senior school for summative assessment.

Fig. 4.3 Science-related
ideas used by the students

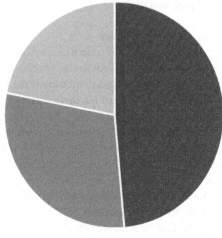

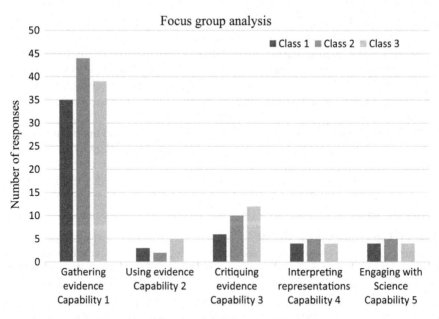

Fig. 4.4 Science capability-related responses made by the students

These are Not Achieved; Achieved; Merit and Excellence. They were converted to percentages and are presented in Fig. 4.5.

The results showed some evidence of student learning during the course of the topic students had studied and the intended content and aspects of planning and critiquing an investigation presented to them as a scenario. We acknowledge that this evidence is weak as we do not have pre and post data and we are unable to comment whether the students who have not achieved were present in class when the information was taught.

Results of Student Questionnaire
To elicit students' thoughts about their understandings about the purpose of school science and their ideas about science investigation, a questionnaire was administered at the end of the unit on water. Students were asked the following:

- What activities they did in their science class?
- Why do they think they did these activities?
- How did they know what they were doing is science? and
- If there were no investigations, what they would learn in science?

Fifty-seven students completed the questionnaire. More students said that the purpose of science investigation was for them to further their knowledge and some said it was to explain things and to understand. For example, one student in Chris's class wrote:

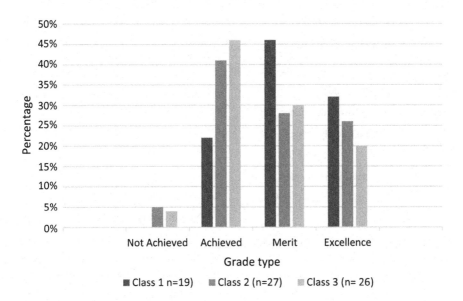

Fig. 4.5 Student grades in end of topic test

Investigations help us to answer our questions…and to find out if something will work.

A student in Fran's class suggested (Fig. 4.6):

…so we can explain the data we collect and compare it with other groups.

For types of activities, the codes were generated from student responses. For example, Student A in Hazel's class said:

Well, we do practical in most classes, collect our data, then based on that we write our explanation or theory.

We coded this as two responses: Practical, and Explanation/theory.

When asked about the science-related activities they engage in, more student responses said they did testing and experiments ($n = 68$) and nearly half the responses ($n = 28$) indicated science investigation and making observations. Twenty responses indicated that they were asked to offer explanations and their own theories ($n = 20$) (Fig. 4.7).

To further understand what students thought about the role of investigation in their science learning, they were also asked what they would learn if there were no science investigations. More said they would learn something, but that it would be boring ($n = 8$). Three thought they would learn as much as the teacher knows. Survey responses have their limitation in that we do not know what these students understood an investigation to be.

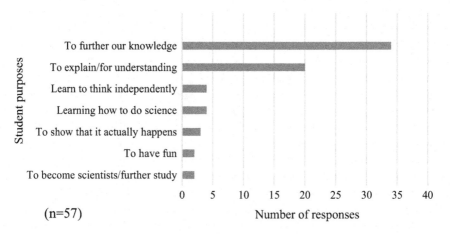

Fig. 4.6 Students' views about the purposes of science investigation

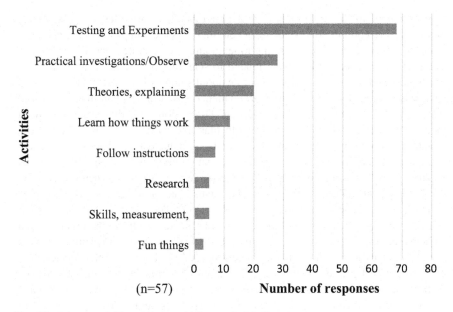

Fig. 4.7 Science activities students said they undertake in science

4.4 Summary

Teacher beliefs about science investigation and why students ought to engage in it were reflected in teacher practice. All three teachers believed in providing opportunities for their students to engage in science investigation and this was reflected in their planning and in all observed lessons. The literature suggests that teachers do practical work for motivational reasons (Abrahams & Millar, 2008), and students do not always learn what the teachers intend them to learn; this case study provides some evidence of learning from doing the investigations.

All three teachers gave students learning opportunities that covered all three educative purposes for school science investigation, that is, developing investigative skills, substantive understanding and developing an understanding about NOS and science capabilities (Figs. 4.3 and 4.4). Teachers used investigation to illustrate science ideas, for example, change of state, developing skills of using equipment correctly, and investigating questions that they wanted to find answers to. The content taught and learnt was as Millar (2004) puts it, to bring '"students" understandings closer to those of the scientific community' (p. 2) and aligns closely with one aim of science which is for students to understand the established body of scientific knowledge at an appropriate level for students' needs, capabilities, and interests (Millar, 2004).

Although students learnt about the investigative process, it was not always recipe practicals that they engaged in. Hipkins et al. (2002) suggest that such investigations are common in New Zealand schools. The participating students planned investigations, made observations, gathered evidence and offered explanations. Teachers were committed to students developing understandings about the nature of science, and students were beginning to learn and demonstrate science capabilities. Their nature of science ideas were related to scientific practice and not rote-learnt tenets of the nature of science. It was more about students learning that they have to have robust data, and their theories are evidence-based rather than their being told that *scientific theories are evidence-based*. Discussion with the teachers showed their strong desire for students to learn to plan and carry out investigations from which they could draw valid and evidence-based conclusions. The three teachers had different approaches to investigation that aligned with their beliefs about whether it was for students to learn to investigate in a structured manner, or to be given more control on what they wanted to investigate, or to use everyday contexts to make sense of science in their everyday lives. The practice of having clear learning intentions at the start of the lesson and reflection to finish the lesson so that students left the class knowing what they had learnt from their investigation were useful practices to enhance students' learning. Although having intended learning from one investigation, not all learning intentions could be met in each lesson. Abrahams and Millar (2008) strongly recommend that teachers should aim to have fewer learning intentions, share them with the students, and follow-up to ascertain they have been learnt. Similarly, Hodson (2014) argues that the learning approach needs to be closely related to intended learning. The following statement exemplifies Fran's understanding of Hodson's idea:

> You don't just do practical work or investigation, I think about what I want the students to learn, and then decide what might be the most appropriate way for the students to learn that…often it is an experiment, sometimes an investigation where they are looking to confirm a theory we have learnt about, at other times they may be learning the procedure and skills. I might want them to gather evidence or I might present some evidence… graphs, pictures… and want them to critique it.

Consider these

1. What purposes do you have for including science investigation in your teaching?
2. What might be some strategies practiced by these teachers practice that may be pertinent to addressing purposes you have identified?
3. The chapter argues that there is a close relationship between teachers' beliefs about science investigation and how they teach it. Consider your own beliefs about why students should learn to investigate, and critically reflect on how you teach or would teach science investigation.

References

Abrahams, I. (2011). *Practical work in secondary science: A minds-on approach*. UK: Bloomsbury Academic.

Abrahams, I., & Millar, R. (2008). Does practical work really work? A study of the effectiveness of practical work as a teaching and learning method in school science. *International Journal of Science Education, 30*(14), 1945–1969.

Anderson, D. (2015). The nature and influence of teacher beliefs and knowledge on the science teaching practice of three generalist New Zealand primary teachers. *Research in Science Education, 45*(3), 395–423.

Fernandez, T., Ritchie, G., & Barker, M. (2008). A sociocultural analysis of mandated curriculum change: The implementation of a new senior physics curriculum in New Zealand schools. *Journal of Curriculum Studies, 40*(2), 187–213.

Hipkins, R., Bolstad, R., Baker, R., Jones, A., Barker, M., Bell, B., et al. (2002). *Curriculum, learning and effective pedagogy: A literature review in science education*. Wellington, NZ: Ministry of Education.

Hodson, D. (1998). *Teaching and learning science: Towards a personalized approach*. UK: McGraw-Hill Education.

Hodson, D. (2014). Learning science, learning about science, doing science: Different goals demand different learning methods. *International Journal of Science Education, 36*(15), 2534–2553.

Hume, A., & Coll, R. (2010). Authentic student inquiry: The mismatch between the intended curriculum and the student-experienced curriculum. *Research in Science & Technological Education, 28*(1), 43–62.

Lowe, B., & Appleton, K. (2015). Surviving the implementation of a new science curriculum. *Research in Science Education, 45*(6), 841–866.

Millar, R. (2004). The role of practical work in the teaching and learning of science. *High school science laboratories: Role and vision*. Retrieved from http://www.informalscience.org/images/research/Robin_Millar_Final_Paper.pdf.

Ministry of Education NZ. (2007). *The New Zealand Curriculum*. Wellington, NZ: Learning Media.

Chapter 5
Science Investigation in Primary School: Changes to Teacher Practice

5.1 Research-Informed Reflection

One aim of science education research is to inform teaching with a view to improving learning outcomes for students. Policymakers emphasise the need for evidence-informed practice. Our experience suggests that teachers continually reflect on and review evidence from their own practice with a view to improvement. We were curious to know what happens when published research evidence is added to teachers' reflection processes. We report the research-informed changes that teachers made to their practice of school science investigation and how these changes influenced students' learning. In this chapter, we present the two primary school cases; Chap. 6 presents the case of a secondary school. In both chapters, we describe the teachers' intentions, the ensuing changes to practice and the student learning that resulted.

As described in Chap. 2, a summary of initial findings was presented to the teachers as a group. The data were summarised uncritically—we described what we saw, not what we felt were omissions or improvements needed—in terms of practice and student learning. We shared the range of approaches to investigation that we had observed across contexts, examples of practice, and a summary across the contexts of students' perceptions of the investigations they engaged in and the purposes they saw for them. We noted that there was much to celebrate in their existing practice with regard to science investigation. The teachers had time for collegial discussion about the findings and the challenges and successes experienced in their practice. They had the long summer break in which to reflect on their practice and engage with the literature provided (see below) about improving learning from science investigation. Early in the new school year, each researcher met with their respective teachers, individually in the case of the primary teachers, supporting them in planning to implement the changes they wanted to make. The teachers were responsible for identifying these changes, making planning decisions, and then implementing the changes.

© Springer Nature Singapore Pte Ltd. 2018

A. Moeed and D. Anderson, *Learning through School Science Investigation*,

https://doi.org/10.1007/978-981-13-1616-6_5

5.2 Primary Teachers' Intended Changes

The two primary teachers each chose a different aspect of practice to work on. Alison felt that although she had selected the Bubbles topic based on student engagement with the topic, and because there was a strong potential for student-led investigations appropriate to their level, the substantive learning involved was difficult and too abstract for young students. She decided that she wanted students to know why they were learning about a science topic and to see its application, developing the 'Engage with Science' capability (Bull, 2015). This focus also aligned with the New Zealand Curriculum achievement objective for the Nature of Science Participating and Contributing strand for Level 1/2: 'Explore and act on issues and questions that link their science learning to their daily living' and the Physical World contextual achievement objective: 'Explore everyday examples of physical phenomena, such as movement, forces, electricity and magnetism, light, sound, waves, and heat' (Ministry of Education, 2007). The researcher connected her with the article *Principles and Big Ideas of Science Education* (Harlen, 2010) to help inform her choice of topic and associated substantive focus ideas. Alison independently planned a topic on magnets because she felt the properties were easily observable and children could recognise the ways magnetic properties can be used. She wanted her students to be able to describe how a magnet behaved, but also to explain why it was important to know about magnets.

Prior to the second phase of the research, Patsy moved to another school in a higher socio-economic area and was teaching a younger class than the one she had taught in Phase 1. For the purposes of the research, she taught a class, not her own, of the same age group as her previous class (age 9–10). When reflecting on Phase 1, she felt she did a lot of talking in the chemical reactions unit: 'I'd got into that habit of talking too much and not getting them to talk and share and things.' She wanted to engage with students' ideas more through interactive discussion. The researcher supplied and discussed with her two research articles about promoting and managing classroom talk (Chapin & O'Connor, 2007; Stein, Engle, Smith, & Hughes, 2008) to help inform her practice in this area. They discussed these readings together and what they might look like in terms of science teaching practice.

Patsy decided she wanted to investigate heat transfer for the observed unit. The researcher provided teacher resources to support this topic (Ministry of Education, 2001, 2002). Together they planned a unit on heat and insulation which involved students working in groups designing two investigations. The first was a more scaffolded investigation where students were to decide on, and test conditions for, preventing an ice cube from melting, based on a science concept cartoon activity about a snowman melting (Keogh, Naylor, & Downing, 2003). This was followed in session two by a teacher-led class investigation about heat conduction of different materials, where large spoons made of different materials were placed in a pot of boiling water with the handles sticking out. A small knob of butter was placed on each spoon handle and students predicted and observed which knob of butter would melt more quickly. The students then planned another investigation, which was more independently student-owned, and carried this out in the third session. The goal was to keep a cup of coffee warm as long as possible. Both student investigations were carried out in small groups. The unit focused on the Level 3 Nature of Science Investigating in Science achievement objective from the New Zealand Curriculum: 'Ask questions, find evidence, explore simple models, and carry out appropriate investigations to develop simple explanations' (Ministry of Education, 2007). Patsy wanted her students to consider elements of fairness, and to take and record measurements, in designing and carrying out an investigation. She wanted them to make inferences based on their data, building the capability 'gathering and interpreting data'. The substantive ideas Patsy wanted to address through investigations were that heat is a form of energy, that heat can move, and that different materials conduct heat at different rates. These ideas relate to the Physical World achievement objective from the New Zealand curriculum: 'Explore, describe, and represent patterns and trends for everyday examples of physical phenomena, such as … heat' (Ministry of Education, 2007).

5.3 Observed Changes to Primary Teachers' Practice

In Phase 2, Alison maintained many of the features that characterised her practice of science investigation in Phase 1. Again, there were multiple opportunities and dedicated time was given for individual play and exploration. She had enough equipment to allow individual play and exploration—in the initial session all students had a wand magnet and roamed the classroom exploring with it, checking out what it would stick to, and investigating other impromptu student-initiated questions such as how many number fans (a mathematical aid) would stick to one magnet. She again modelled wondering and used this to suggest new aspects to investigate, often making public a student's observation: 'Sean noticed that his magnet pushed Keigan's magnet away. I wonder if anyone else's magnet can push away as well as stick to things?'

As in Phase 1, Alison introduced different equipment to promote different investigations. In the second session, she used timed group circuits of tables with different sets of equipment, each designed to support practical experience with a specific substantive focus. For instance, children sorted a range of materials into piles of sticking and non-sticking to magnets to help identify the nature of materials that are attracted to magnets. They explored whether magnetic force could pass through materials by playing with sealed plastic jars of paper clips. Students investigated whether all metals stuck to magnets by playing with a magnet and a container of different types of coins. Alison celebrated and made public children's observations during exploration, but also more formally through class reflection times on the mat following explorations: 'What did you notice about magnets?' She encouraged careful observation beyond just seeing: 'What did you feel?' and encouraged children to use evidence from their observations: 'How could we check if it is a magnet?' She introduced disconfirming evidence that challenged children's thinking by bringing in a clip that stuck to metal surfaces through suction. As in Phase 1, Alison managed classroom dialogue through a mix of interactive approaches (Mortimer & Scott, 2003), using a more authoritative approach to make substantive concepts about magnets explicit, e.g. 'Do all metal things stick to magnets?' She continued to describe children as scientists and linked what the children were doing to science: 'What makes us good scientists?' and using the scientist poster again—one each depicting a boy and a girl: 'What are your scientist's eyes for?'

A major difference observed in Alison's practice in Phase 2 was a greater focus on developing substantive ideas. As described above, Alison deliberately selected equipment for children to explore, and asked prompting questions that would lead to productive substantive ideas about magnets (see Fig. 5.1a, b and c).

This change in emphasis came at the expense of Alison's focus, apparent in Phase 1, on developing and answering children's own questions through school science investigation. The investigations in Phase 2 were more obviously teacher selected and initiated for a substantive purpose, although students owned the investigations through their play and exploration with the equipment that was provided. They took the investigations in directions that caught their attention and interest.

Another clear addition from Phase 1 was the connections Alison made to everyday life. The unit was introduced through a technology opportunity to make a magnet to attach the children's spelling 'rockets' (list of words they were learning) to their fridge at home. In the first session, as well as exploring magnets in their own classroom, the students were taken to the school office where the school secretary showed them the different ways she used magnets in her work—whiteboard magnets, a paperclip holder and a fridge magnet. For homework children drew the magnets that were used in their homes. The parent of one child who used an electromagnet in their business brought it in to show the children how strong it was, lifting a chair (the children were very impressed!).

The final part of the unit was devoted to making and testing the fitness-for-purpose of their own spelling rocket magnet. The connections between the science the children were experiencing at school and everyday life and work were made obvious—Alison's intention in selecting this topic was explicitly enacted in her practice.

As described earlier, Patsy wanted to decrease teacher talk and increase the amount of student participation in science-related discourse. Her situation, working with a class that was not her own, made this goal more of a challenge as she did not know the students and their names well. However, she made strong attempts to increase class discussion and help students to make inferences from what they were seeing as we will see from the evidence that follows. In the first session, students worked in groups to investigate which materials would slow down the melting of an ice cube. In the second session, students observed a demonstration of heat conduction through different materials, using different long-handled spoons in boiling water and measuring the time taken for butter on the end of each to melt.

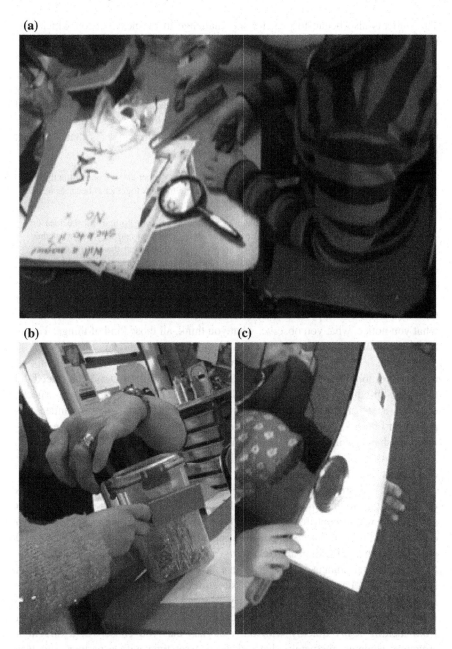

Fig. 5.1 a 'This isn't sticking'—a range of different materials focused students' observation on those materials that are attracted to magnets and those that are not. **b** 'The paper clips are moving up too!'—exploring whether a magnetic field can pass through materials using a container of paper clips. **c** 'Look! It sticks with three pieces!'—investigating the strength of a magnetic field with layers of cardboard and a metal lid

The students also began to plan for the challenge in the next session which was using what they had found out to slow down the heat loss from a cup of coffee.

Like Alison, Patsy maintained practices observed in Phase 1. She continued to emphasise the importance of careful and rigorous observation and link this practice to science, making the nature of science explicit: 'one of the things we know very well is that observation is really important. Science is about explaining the world around us and the way you do that is by taking care, looking very carefully at the things that you're investigating.' She continued to highlight ways that scientists work, linking it to what was expected of the students:

> One of the things that scientists have to do is a lot is work together and it's one of the things we're also going to do today and the reason we're going to work together is because we're going to pool our information, our data, the things we find out.

> Often what scientists will do is they have an idea but they can't go and test the idea on actual big stuff so they make a model of it, like a small version and they use what they discover about that small version of things to give them ideas about the big thing. So we haven't got a snow man, so we are going to use ice cubes. Same principle, okay, fits in the room, not so messy.

Patsy used a similar approach to introduce her expectations concerning student talk: 'I'd be really happy to hear lots of talking to each other about science, about what you notice, what you observe, what you think, all those kind of things. That is really helpful and useful and that's how scientists figure things out.'

In Phase 1 Patsy began most sessions with a long period of talk—a mix of teacher questioning and teacher talk. In the first lesson, this talk time introduced the topic; in subsequent lessons, it recapped the previous session and clarified aspects, e.g. that acids and carbonates react, not liquids and solids. She continued this pattern in Phase 2, but obvious differences were noted in the nature of the opportunities she provided for student talk and elicitation of student thinking. A comparison of the nature and number of teacher questions used during the mat-time at the start of the first session in each phase is presented in Table 5.1.

Table 5.1 illustrates the way in which the amount as well as the nature and focus of Patsy's questioning changed during this unit. There was a high expectation that all students would contribute, not just those with their hands up, which had been common in Phase 1. She elicited more ideas from students more frequently and engaged them in thinking and reasoning about each other's science ideas. This focus on greater student involvement in talk continued at the start of the second lesson in which Patsy used a doughnut structure for students to share their findings from the previous session: 'Now what you're talking about is yesterday's experiment—what you noticed about that ice cube melting and what you think stopped it from melting really quickly with the materials that you chose.' In a doughnut structure, students alternately share ideas in a set time with a partner who was facing them in a circle formation. One circle is then moved clockwise or anticlockwise to allow a new partnership to share ideas. Patsy then called on different students to share what they had been talking about.

Table 5.1 Comparison of Patsy's questions during the opening session from phase One and phase Two

	Phase 1	Phase 2
Duration of opening teacher-led discussion	15 min and 43 s	15 min and 21 s
Number of teacher questions to students	9	17
Interaction type generated by questions	Inquiry/response/evaluation (IRE): 5 (instant student response from student with hand up) Think/pair share (TPS): 1 (30 s, part way through teacher questioning) Contribute to brainstorm: 1 (3 min-recording ideas of students with hands up) Show of hands: 2	Think pair share (TPS): 1 (1 min, and used before any questions asked) Eliciting student ideas (EI) (named student, no hands up): 9 Eliciting ideas (general): 1 Requesting add-ons (AO) (named student) to another student's idea: 3 Eliciting student's reasoning (ER): 5 Connecting students' ideas (CI): 1
Examples of teacher questions	What's an acid? (IRE) Can you think what made it a chemical change? (IRE) What is chemistry? (TPS and Brainstorm-individual responses)	What kinds of things could we use to help us slow down melting? (TPS) Johnny, what kind of things did your group talk about? (EI) Can you add something to that, Dominic? (AO) How are you thinking that will help? (ER) Harriet you disagreed there and that's cool...What was your thinking that was different maybe to that? (ER) That links in a bit with the idea Billie had, didn't it, about the pencil case because that's an enclosed space, as well. (CI)

Opportunities for student talk in all three sessions allowed students to consider both substantive understanding and investigative ideas. Much longer periods of discussion were observed in Phase 2 where Patsy involved all students in a focused discussion of the substantive aspects of the findings from investigations and what they may suggest about heat and insulation. For example, in class discussion at the end of the first session, Patsy asked: 'What I want you to think about is—the things that melted really quickly—what did they have in common?' Students discussed this question in small groups, then fed back to the whole group. The doughnut activity at the start of the second session similarly reviewed findings from the ice cube melting activity: 'So what we're interested in then, in that experiment, is the

things you chose to use. How effective do you think they might've been and why?'
The students considered substantive ideas further in a full class discussion following the doughnut activity:

Harry	I wish it was like a bigger box because it would be more stuffing and there wouldn't be as much heat.
Patsy	Okay. So you think the size of the box had an effect. What about you, Izzy, what did you think?
Izzy	If it was a plastic box, it would be better because then it might not melt so fast.
Patsy	Oh, okay. So it would be more useful if it had been a different material. That's an interesting idea. Let's just think about that idea, the fact it was made of cardboard…Do you think the fact it's made of thick card had something to do with the heat, the way the heat worked? Anyone else have an idea?
Billie	Well cardboard might act as insulation, so it might melt quicker in a metal box.
Sonali	It might keep it cooler?

Patsy actively encouraged students to consider where the heat may have come from to melt the ice cubes:

Patsy	When you did the melting things, who was in the group that ran around, shaking it? Something did melt it. It just took a while. Where was the heat there? Where was the thing that melted it, do you think?
Tom	Umm, well maybe…humans, if they run, if they exercise, then they get quite hot, so maybe it's the same with the ice cube – like they're moving around a lot – they like get hot.
Patsy	Yeah, yeah, good thinking.
Billie	Or you're shaking it around – some of it comes off?

Discussions also focussed on investigative processes. During the teacher-led spoon investigation, Patsy raised ideas about rigorous observation and the students noticed aspects that were unfair, which led to a discussion of investigative ideas about fairness and control:

Patsy	You couldn't really, could you really look at all of them at once and do this well?
Several students	No.
Patsy	You couldn't, could you? You'd have to carefully look at one….
Sarah	Yeah, but there's also one thing as well.
Patsy	What's that?
Sarah	Umm, they're not the same size because the plastic one's more closer to the steam.
Patsy	It is, you're right. Yes, that's right.
Sarah	So it's not fair.
Patsy	It's not fair, okay

Patsy then encouraged the students to identify ways to make the investigation fairer, concluding: 'So there's quite a few things that if you wanted to make sure that this was a really accurate thing, there's quite a few things you'd have to have the same, wouldn't you, to get a really good, accurate result.'

In comparing Phases 1 and 2, the second major difference observed in Patsy's practice was that the students, working in small groups, had more ownership of the investigations. In the melting ice cube investigation, Patsy provided a range of materials such as tin foil, sawdust, fabric and bubble wrap, as well as two choices of box to keep them in. Students were required to make collaborative and reasoned choices in their groups about these materials. The rest of the investigation was scaffolded in that the students all did it at the same time, with Patsy governing the regular intervals at which they recorded observations of their ice cubes. In the final investigation the students, working in their groups again, were responsible for planning the way in which they would keep the coffee warm and identifying which materials and conditions they would use. Patsy kept the design simple and manageable by only requiring that the temperature be taken at the start and end of a common 20-min interval.

5.4   Impact of Changes of Primary Students' Learning

In Alison's class, the second set of observations, the following year, confirmed the tentative findings of the first—that children most often described easily observed patterns as their learning, e.g. magnets stick to (some) metals. When asked how they knew, the students interviewed most often referred to an observation they had made during practical explorations. This second unit focused intentionally on more easily observed patterns and phenomena, but also on the application of their knowledge—what magnets could do and why it was important to know about magnets.

To assess students' learning from the unit, before and after the lesson sequence each child was asked three question and their responses were noted. The three questions were: (1) What is a magnet? (2) What does it do? (3) Why is it important to know about magnets? Tables 5.2 and 5.3 present the analysis of the data.

Table 5.2 shows that for substantive ideas, while most students knew before the unit that magnets stick to metals, all knew this by the end, with a large proportion additionally recognising that not all metals are attracted to magnets. New understandings included that magnets could repel each other. Many thought that magnets were made of metal at the end of the unit, perhaps confusing what they were sticking to with the magnet itself, or perhaps concluding that the dark metallic look of the magnet made it metallic. The substantive ideas that students came to

Table 5.2 Substantive learning about properties of magnets by new entrant students

Student understandings about magnets	No. of student responses prior to unit ($N = 18$)	No. of student responses after unit ($N = 17$)
Magnets stick to metals	14	17
Not all metals stick to magnets	0	11
Coins that don't stick to magnets are not metals	0	1
Magnets stick to things	6	1 (non-metals)
Magnets are made of metal	3	10
Magnets stick to each other	4	1
Magnets can repel	0	3
Other properties (e.g. hard, have a shape)	3	2
Awareness of invisible force	2 ('They are sticky but you can't see the sticky' 'it is like glue you can't see')	3 ('It can go through stuff' 'it sticks to metal with a field you can't see' 'It goes through things without even touching them')
Some magnets can be stronger than others (stronger magnets hold heavier things)	0	1

Table 5.3 New entrant students' views about the application of learning about magnets why they should learn about them

Students' perceptions of what magnets do	No. of student responses prior to unit ($N = 18$)	No. of student responses after unit ($N = 17$)
Pull things	1	
For play	3	
They can hurt you	1	
Utilitarian purpose—to hold/stick/connect	7	13

Why students thought they should learn about magnets	No. of student responses prior to unit ($N = 18$)	No. of student responses after unit ($N = 17$)
To know things	5	5
To teach others	3	5
To know what magnets do—how to use them	7	8
In case you want to be a scientist	1	
Because it is science	1	
Not sure	1	1

understand through the unit were easily observed, so more easily developed through observation than the more abstract substantive ideas related to the bubble unit in Phase 1 about what happens when balloons pop.

In considering the investigative skills of Alison's young students, similar observations were made to Phase 1. When the equipment was restricted by the teacher to guide the investigation or when they were directed to observe something immediately observable, the students were able to make specific observations which they could use to answer closed questions set by the teacher. When left to play freely with equipment, the students followed their own lines of inquiry to answer their own questions, a few demonstrating a logical approach and perseverance. For example, two boys worked in a pair to add number fans on a metal clip to a magnet, one at a time until some started to fall off, just to see how many it could hold. Another girl purposefully added layers of cardboard between a lid and a magnet to see how many it could pass through and still stick.

An important enhancement for Alison in Phase 2 was that children's science learning should be relevant and applicable for them. This aspect addresses the objective for the Participating and Contributing aspect of the Nature of Science strand for Levels 1 and 2 of the New Zealand Curriculum for students to link their science learning to their daily lives (Ministry of Education, 2007). Table 5.3 presents findings with regard to this aspect.

All students could identify uses and applications for magnets and these views were more consolidated on connecting, sticking and holding things around the home and classroom by the end of the unit. Alison shared later that the students used their fridge magnets to hold up their spelling rockets as she had planned, but then independently found many opportunities to use them in other ways, e.g. to secure newspaper items and homemade jigsaw pieces to the teacher's whiteboard or as hooks to display lost pairs of swimming goggles. While students most commonly saw a utilitarian purpose for learning about magnets—to know what they do—the number responding in this way was unchanged from the start of the unit, so Alison's goal was not really attained by the majority of students. For these students, the most common reason was simply to know things—they seemed to think that knowing things was the natural outcome and reason for being at school. Several thought that you learnt about things such as magnets in order to teach others about them. One student interviewed at the start of the unit seemed to recognise that investigating curious phenomena was a feature of science: 'Because it's good science because something happens—but I'm not sure what!' This comment is unusual as students mostly struggled to associate what they were doing with science or describe what they learnt about science itself. In hindsight, for this aspect for young children it may have been useful to ask something like: 'Do you think you behaved like a scientist today?' or 'What did you do that was like a scientist?' instead of asking what they learned about science.

In Patsy's class, a questionnaire (similar to the one in Appendix 3.2) was used to probe students' perceptions of intended and actual learning from the investigations they had experienced. Analysis of the responses are shown in Tables 5.4 and 5.5.

Table 5.4 Year 5/6 students' perceptions of what Patsy wanted them to learn from the heat conduction investigations

What do you think Patsy wanted you to learn from doing the science activities?	N = 18
About heat/heat energy	9
Results of investigation/about keeping something cold/hot	6
Different ways to learn about heat (investigation processes?)	1
To see our point of view (was aware that she wanted their ideas)	1
About degrees (temperature)	1
To know stuff	1
No response	1

Students' responses to this question show that they mostly thought that Patsy wanted them to learn about heat; all responses were focused on substantive aspects. In Phase 1, while many students (10/26) suggested substantive ideas about reactions, the same number also suggested more syntactic ideas about how scientists or science works, perseverance and careful observation. Syntactic ideas were not as common in Phase 2, suggesting the greater focus on substantive ideas in their class and group discussions perhaps led students to think this aspect was Patsy's intended learning focus, despite her inclusion of discussions about fairness and her many comments linking their work to that of scientists.

Students' substantive responses to the post-unit questionnaire are shown in Table 5.5.

These responses show that nearly all students were making their own sense of the investigations. Some were developing productive and scientifically accepted ideas about heat and insulating properties of materials. Several also held ideas that were quite different from the scientific explanations for the phenomena they had been investigating. This mixture of ideas is also indicated in students' comments in class discussion. Tom's comment, reported earlier, that associated humans getting warm through activity with ice cubes that were shaken melting faster is an example of a potentially fruitful idea. Similarly, Billie's idea that an ice cube would have melted faster in a metal box shows a developing understanding of heat conduction properties. Alternative conceptions were also displayed such as Sonali's response that a metal box may have kept the ice cube cooler. Patsy allowed these ideas to sit equally alongside more accurate ideas. Whereas Alison manipulated the discourse with her younger students by ignoring incorrect ideas and publicising more accurate conceptions, it seemed Patsy needed a more sophisticated range of strategies to address alternative conceptions with her older students. Some ideas, such as the metal keeping the ice block cooler for longer, could well have been addressed with further investigations, but time was short. Being able to manage discussions that guided students' collective thinking toward scientifically acceptable ideas required more than the techniques that Patsy had mastered for eliciting students' ideas and getting them to engage with ideas of other students as she recognised in her reflections after the unit:

Table 5.5 Year 5/6 students' substantive learning from investigations

When we tried to keep our ice cubes as long as possible they still melted. What do you think made them melt?	N = 18
Heat that was present (there was at least some heat that made it melt)	6
The surrounding material (e.g. tinfoil, sawdust)	4
The air	2
Time (e.g. we left it overnight)	4
Not enough materials	2
Lack of really cold air	1
Hot air got trapped and melted it	1
Classroom	1
No response	1

What do you think happened to the heat from the coffee cup you were trying to keep hot?	
Heat tried to escape/escaped Two students included additional comments: … because not enough insulation (1) … but our insulation slowed it down (1)	8
Heat stayed there Two students included additional comments: … there was no means of escape (1) … it was stored and was given out like a solar panel (1)	4
Maybe it got stuck inside the box and stayed there	1
The air pressure kept it there/air was too strong	2
It evaporated	1
Environment	1
Don't know	1
No response	1

What kind of materials do you think would help keep the coffee hot?	
Batts (roof insulation)	11
Heater/stove/hair dryer	**9**
Tinfoil	8
Fleece	5
Wool	5
Fluffy stuff	3
Clothes	3
Warm things	3
Mug warmer	3
Dark colours	2
Blankets	1
Bubble wrap	1
Cotton	1
Hot air	1

(continued)

Table 5.5 (continued)

How do you think these materials would help keep the coffee hot?	
How do you think these materials would help keep the coffee hot? Property of the materials—they keep it warm—it is what they do?	8
They have their own warmth	2
Hot things add heat to keep it warm (hair drier, etc.)	2
They keep heat in	1
Absorbs heat	1
Stores heat and gives it out like a solar panel	1
Describes how to wrap in insulation materials	1
No response	2

> I liked the approach of asking everyone not just those who put their hand up as it was a science community figuring out things together. I want to keep working on that approach… Now I will try to record the students' responses and observations more and refer back to them- kind of build up their understanding a bit more purposefully – point out an aspect they noticed and direct them to use that to think about something else.

In addition to the post-unit questionnaire, students' substantive learning about insulation and heat transfer was examined before and after the unit using a concept cartoon activity about the effect of putting a coat on a snowman (Keogh et al., 2003). The post-unit responses were added in a different colour to their earlier responses. This assessment was more challenging than the questionnaire and probed students' understanding of the role of insulation. Results showed that most students held misconceptions at the start of the unit that the coat would melt the snowman because it would make him warm, and there was little change at the end. Only four students' final responses showed greater understanding about heat than prior to the unit.

Many scientific explanations, such as those about heat, are complex, abstract, and frequently counter-intuitive. Most children tended to rely on their common experience of a coat making them feel warmer, or stopping the cold air getting in, as in closing a draughty window, for example. None of them transferred their observations about the role of insulating materials in slowing down the melting of an ice cube or any learning gained through that experience and discussion to address all the ideas expressed in the concept cartoon, although some included more scientifically sound reasons for some responses. These complex ideas take much more development than a series of pertinent investigations and elicitation of student-generated explanations allow.

With regard to investigative skills and understanding about science itself, Patsy included investigative ideas as a focus for discussion in the second session, and some students spontaneously identified unfair aspects of Patsy's heat conduction demonstration, as reported earlier in the chapter. In the post-unit questionnaire, the most common responses about behaving like a scientist were: 'Try to make it fair test' (5), 'Persevere/don't give up/keep trying' (5), and 'Share ideas' (4); all these were aspects emphasised by Patsy. However, similar numbers also included aspirational comments such as 'Believe in yourself' (4) which did not feature in observed discussions. Other responses included: 'Be curious' (2) and ideas about integrity: 'Don't cheat'/'tell the truth' (3), and 'Don't always expect something amazing to happen'. Some less scientific ideas included: 'Listen to your teacher' and stereotypical views of science 'Be crazy' and 'like explosions' (2). Most students appeared to be developing worthwhile ideas about what was involved in science and scientific investigation; teacher emphasis appeared to play a role in the development of these ideas.

5.5 Summary

What can we learn from these observations of primary teachers making changes to refine and enhance their practice in order to improve student learning during science investigation? First, that given adequate opportunity, time, data and information on which to reflect, these teachers were able to identify an aspect they wanted to enhance that was relevant, worthwhile, and specific to their practice. Although, as researchers, we had not specifically targeted any aspects of the teachers' practices in discussions with the teachers, we agreed that the aspects teachers identified were those where enhancements could be made. Second, the teachers drew on targeted support in order to make changes to their practice—they engaged with readings provided by the researchers and discussed with them ways they would implement ideas they had identified in their practice.

In terms of the nature and impact of changes made, both teachers made observable changes to their practice that influenced their students' learning, particularly in relation to substantive science ideas. The students in Alison's class developed new understandings about the behaviour of magnets and related that learning in their everyday lives. Patsy changed the nature of the classroom discourse so that more student ideas were elicited and discussed. Students had an opportunity and were able to suggest possible explanations for the observations they made during investigations. Some students demonstrated more scientifically accurate substantive understanding; however, many misconceptions about heat remained.

Looking across both cases and phases it seems that substantive learning that is easily observable and directly inferred from science investigations is more readily developed by younger students. Students' learning through exploration and play can be guided towards these ideas through selecting appropriate equipment and targeted teacher questioning. Teacher-guided discussion can easily focus students on relevant observations which they can use with appropriate prompting to make useful inferences and develop fruitful substantive understanding. Science concepts that are more abstract and involve less intuitive inference need more specific and planned support. Such support may include further investigations that allow students to defend or refute their suggested explanations with evidence. Teacher discourse strategies are needed that introduce accurate substantive science ideas for students' consideration, as well as challenge less fruitful ideas through debate in the light of evidence from their experiences and investigations.

In considering the development of investigative skills and understanding of the nature of science, the findings from this chapter support those from chapter three in that when teachers are explicit in modelling, identifying and describing scientific behaviours, students begin to adopt them for themselves and associate them with science.

Consider these

1. How can teachers make scientific behaviours and practices more explicit when scaffolding students' engagement in the practice of science investigation?
2. Which substantive ideas that form part of your classroom science curriculum are more abstract and difficult for students to develop? Which further investigations could you include that would provide more empirical evidence for students and challenge alternative conceptions? Which discourse strategies could you apply to better support students with their development?

References

Bull, A. (2015). *Capabilities for living and lifelong learning: What's science got to do with it?* Wellington: New Zealand Council for Educational Research. Retrieved from http://www.nzcer.org.nz/system/files/Capabilities%20for%20living%20and%20lifelong%20learning%28v2%29.pdf.

Chapin, S. H., & O'Connor, C. (2007). Academically productive talk: Supporting students' learning in mathematics. In W. G. Martin & M. E. Struthens (Eds.), *The learning of mathematics* (pp. 113–128). Reston, VA: National Council of Teachers of Mathematics.

Harlen, W. (Ed.). (2010). *Principles and big ideas of science education.* Hatfield, UK: Association for Science Education. Retrieved from http://www.interacademies.net/File.aspx?id=25103.

Keogh, B., Naylor, S., & Downing, B. (2003, August). *Children's interactions in the classroom: Argumentation in primary science.* Paper presented at the European Science Education Research Association Conference, Noordwijkerhout, The Netherlands. Retrieved from http://www.conceptcartoons.com/resources/esera-paper.pdf.

Ministry of Education. (2001). *Making better sense of the physical world.* Wellington, NZ: Learning Media.

Ministry of Education. (2002). *Insulation: Keeping heat energy in: Building science concepts book 47.* Wellington, NZ: Learning Media.

Ministry of Education. (2007). *The New Zealand curriculum.* Wellington, NZ: Learning Media.

Mortimer, E. F., & Scott, P. H. (2003). *Meaning making in secondary science classrooms.* Maidenhead: Open University Press.

Stein, M. K., Engle, R. A., Smith, M. S., & Hughes, E. K. (2008). Orchestrating productive mathematical discussions: Five practices for helping teachers move beyond show and tell. *Mathematical Thinking and Learning, 10,* 313–340.

Chapter 6
Science Investigation in Secondary School: Changes to Teacher Practice

6.1 Research-Informed Reflection

Investigating in science is given primacy in New Zealand secondary schools. As mentioned in Chap. 2, a mandated requirement is teaching about the nature of science and students developing investigative skills. Educational policy considers research evidence-based teaching to be desirable and encourages teachers to inquire into their own teaching practice (Ministry of Education, 2007, p. 35). We have observed that just like the primary school teachers in the previous chapter, secondary school science teachers continually make changes to their practice to support student learning more effectively through a process of reflection and review. However, this process does not tend to be informed by published research. As in Chap. 5, our interest was to find out how access to published research evidence added to teachers' reflection and the impacts it had on how they use science investigations in their classes. The extant science education literature highlights that although teachers engage their students in practical work (science investigation being a subset of practical work), there is little evidence that students learn from this engagement. Chapter 5 presented the two primary school case studies and, in this parallel chapter, we describe what happened when participating secondary school teachers made research evidence-based changes to their practice of science investigation.

As reported in Chap. 2, a summary of findings from Phase 1 was presented to the teachers as a group. We described our observations, we did not criticise teacher practice, and did not highlight any possible oversights or point out the improvements that could be made. Our summary showed that all three teachers were providing opportunities for their students to engage in a range of approaches to investigation and that we observed effective science teaching in their practices. Teachers had time to talk about the findings and discuss the challenges and successes experienced in their day-to-day teaching. As with the primary school

© Springer Nature Singapore Pte Ltd. 2018
A. Moeed and D. Anderson, *Learning through School Science Investigation*,
https://doi.org/10.1007/978-981-13-1616-6_6

teachers, our intention was for the three teachers to have time to reflect on our findings during the long summer break and engage with the literature provided about improving learning from science investigation (see Sect. 2.1).

6.2   Changes the Teachers Intended to Make

Early in the New Year, a researcher met with the secondary school teachers as a group and during the discussion that followed the teachers identified the changes they wanted to make. The teachers were very clear about the investigative skills they wanted the students to learn from investigations (e.g. planning, controlling variables). They had started to implement the Science Capabilities for Citizenship designed to support New Zealand science teachers in implementing the Nature of Science (NoS) strand of the curriculum (Fig. 4.4). The findings suggested that their focus of science investigation had been on gathering data through making observations, and then offering an explanation. The findings had also shown that they had multiple learning intentions for each investigation. They talked in some detail about the research they had been reading during the holidays. They had considered the effectiveness of the practical work framework (see Fig. 6.1) proposed by Abrahams and Millar (2008) and decided to focus on fewer learning outcomes for each investigation their students carried out. The intention was to focus beyond engagement (Effectiveness level 1) to what the students actually learn (Effectiveness level 2).

Although the framework was designed by Abrahams and Millar to evaluate the effectiveness of practical work in general, we decided that science investigation being a type of practical work, the framework could be reasonably used for science investigation. An added advantage was that it provided a useful framework for analysing the data we would collect. As a department, the teachers decided to

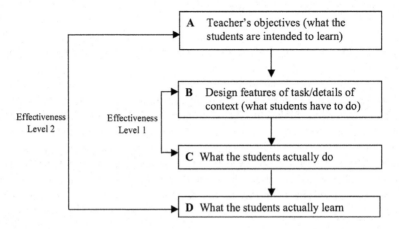

Fig. 6.1 Abrahams and Millar's (2008) framework for analysing practical work (p. 1947)

continue their focus on Science Capabilities for Citizenship. First, the data showed that they had emphasised the science capability of gathering and interpreting data, with a strong emphasis on careful observation and making inferences from these. In Phase 2 they would focus on other capabilities, for example, focussing on evidence-based conclusions and critiquing their planning and critically evaluating evidence. Second, they would focus on Millar's (2010) advice to have fewer and specific learning outcomes for each investigation which they would share with their classes. As a department, they were confident that they could collectively plan to make these changes to their teaching of science investigation in the context of metals and their compounds, which was the next topic they were going to teach.

6.3 Observed Changes to Secondary Teachers' Practice

As stated in Chap. 2, three lessons were observed in each teacher's class, and the teacher's were interviewed at the end of the topic taught during this research. The three lessons taught by each teacher focussed on teaching science capabilities within the context of metals and their compounds. The intention was for the students to be able to plan and carry out a fair testing type of investigation. As they intended, the teachers were seen to be using fewer intended learning outcomes for each investigation and focussed on the science capabilities of using evidence and critiquing evidence. Particularly, they emphasised the importance of accurate measurement, constructing evidence-based explanations, and critiquing the design of the investigation. Although the teachers shared learning intentions with their students in different ways, there was a focus on epistemological understanding in each lesson.

Investigation in Hazel's Class
Hazel took the students through the process of designing and carrying out a fair testing experiment. Students were investigating the effect of temperature on the rate of chemical reactions. This led to students planning their own investigations, gathering data, processing the data and critiquing the evidence. When one group shared their much higher than expected reading, Hazel took the opportunity to talk about dealing with *outliers* in the data. She explained that scientists must look at their data and if there are any measurements that are drastically different then they need to think about why that might be. She encouraged other students to look at their data carefully to identify any outliers, which resulted in the following discourse:

Hazel   Why do you think you have an outlier?
Tom   Maybe we did not read the thermometer correctly.
Hazel   What will you do about this?

Tom asked his group, 'What should we do'? Sarah suggested that they should take one more reading at 40 °C. They repeated two more trials at 40 °C and had readings of 100 s and 104 s:

Hazel   Why did you do three trials?
Sarah   Because doing more trials makes our results more accurate.
Tom   If we did 10 trials it would make it even more reliable.
Hazel   What would a scientist do?
Sarah   Do more trials, like six or something.
David   Depends on how important the experiment is to the scientist. If getting it right is really important or something they may do 10 or even more. If they are just trying to check out that their method works they may do a few, just to check.

Hazel asked the students to critique their alternative plan for repeating the investigation, identifying which variables to control and the best way of measuring and recording. The next two lessons followed a similar style with evidence of students being encouraged to plan and think critically about their design, doing the investigation and critiquing their evidence. Each lesson ended with the class discussing ways of improving the reliability of their findings (Fig. 6.2).

Investigation in Chris's Class

Chris did several investigations, each with a particular focus. For example, in the first lesson, students explored what would happen if they put mentos lollies in fizzy drinks. This experiment illustrates a physical change, although a common misconception is that it is a chemical reaction. A measuring tape was set up outside the building (see Fig. 6.3) and students could put ten mentos lollies into a 1.5 L bottle of a fizzy drink. Chris had set up a measuring tape outside the building, so they

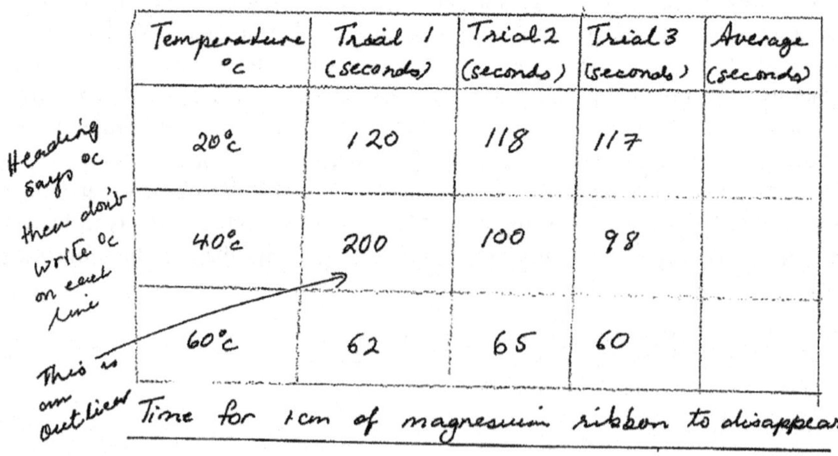

Temperature °C	Trial 1 (seconds)	Trial 2 (seconds)	Trial 3 (seconds)	Average (seconds)
20°C	120	118	117	
40°C	200	100	98	
60°C	62	65	60	

(Handwritten annotations: "Heading says °C then don't write °C on each line"; "This is an outlier")

Time for 1 cm of magnesium ribbon to disappear

Fig. 6.2 An excerpt from Tom's book

could measure the height of the resultant geyser (see Fig. 6.3). It was windy and along with the set ruler students could use a time delay camera to record their measurements. The lesson finished with returning to the class and reflecting on their measurements of the height of the resulting geyser, and what they had recorded. The focus was on the accuracy of measurement. Chris encouraged the students to consider why the data they had collected might not be reliable and they talked about alternative ways of finding out which drink produced the most bubbles. For example, it was windy outside, so the geyser did not go straight up and the readings they were taking were not at eye level. An alternative was set up inside the laboratory (Fig. 6.4). The time delay camera one group used did not work for some reason. There were other barriers to accurate measuring (see Fig. 6.5)!

Fig. 6.3 Student preparing to measure the height of the geyser outdoors

Fig. 6.4 An alternative setup in the laboratory

Fig. 6.5 Another barrier to accuracy of measurement

Fig. 6.6 Simon's group's plan for collecting the displaced gas

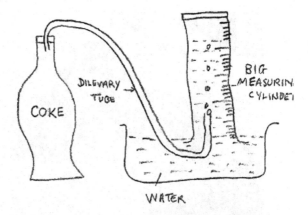

This was followed by students working in groups and thinking about what other measurements they could take that might be more valid and reliable. Tim's group considered measuring the amount of gas escaping from the drink bottle, and Tanya thought they could measure the liquid.

Tim   We could have put a balloon on top of the bottle, then we could see which balloon had most gas.

Tanya   It would be much better to measure the volume of the liquid left in the bottle.

Simon's group drew an elaborate picture of how they could collect the gas by downward displacement of water as they did when they had made oxygen gas. 'We will need a very big measuring cylinder!' (Sue). 'Yeah, and maybe a large bucket with water' (Tony) (see Fig. 6.6).

In the third lesson, in Chris's class, students investigated the effect of temperature on the rate of chemical reactions using magnesium ribbon and dilute hydrochloric acid. Chris showed the class how to use a beaker as a water bath. Students then set up their investigation (see Fig. 6.7). As had happened in Hazel's class, one group found that their rate of reaction was faster for cold water than the warm water. At the end of the activities, the results were collated into a class 'table'. Chris explained that Sarah's group had anomalous results and asked why this might be. Sarah said that they had put the thermometer in the beaker and had not waited for the temperature to warm up the acid.

Chris   So how are you going to sort this in the next lesson?

Sarah   We already repeated it again and put the thermometer in the test tube with the acid and waited. Then we got a better ... more logical result.

Chris   What do you mean by logical?

Sarah   We thought that the magnesium will take less time for the full reaction when the temperature was higher.

Chris   And why do you think that?

Sarah   Heat makes the particles move faster, more collisions ... more useful collisions really. I think you told us about particle theory ...

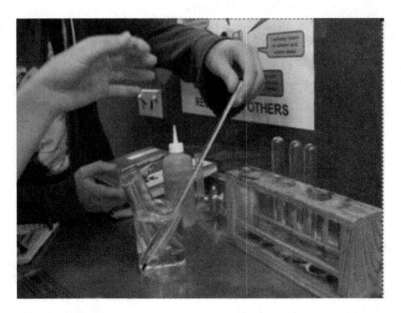

Fig. 6.7 Sarah's group measuring the temperature in the water bath

Investigation in Fran's Class

Fran's class did a similar investigation with mentos and fizzy drinks, but they talked about their plan and how they could collect their evidence so that it was robust before going outside to try it out. The lesson ended with students sharing how they had collected the data and why they thought their findings were reliable and why not. Fran moved around the classroom, asking questions and gaining an insight into the students' understanding on an individual level. Students' ideas were used in reflections at the end of the activity.

Fran often introduced the investigation as a Predict Explain Observe Explain (PEOE) strategy which focussed students to think about their predictions, offer explanations based on their existing ideas, then to observe closely, and construct evidence-based conclusions. With the current focus on using digital technology, students in this school bring their own digital devices to class. Fran uses digital technology in several ways. Sometimes she shows brief clips of an investigation to illustrate a science idea, at other times to start a discussion about a current science-related issue affording the students opportunity to share their ideas amongst themselves and the class. She uses Kahoot (a free multiple-choice learning platform) in a number of ways too, sometimes at the start of the lesson, and at other times as a formative assessment task (see Fig. 6.8).

From time to time, Fran brought newspaper articles or advertisements to class, for example, an advertisement that claimed why a particular antiseptic was best or used an article about the effectiveness of vaccination. Students were encouraged to read the information and look for evidence that supported the claim, being critical about sample size, who was interviewed, were the data representative of the

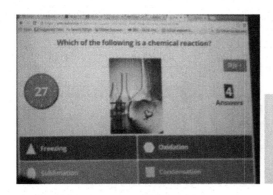
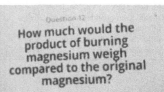

Fig. 6.8 Kahoot tasks used in Fran's class

population or did it reflect the views of only a few? She wanted her students to learn to critique the information presented to them and 'become informed users of scientific claims presented in the media every day.'

Comparison of Teacher Practice Between Phases 1 and 2
All three teachers made changes that they had intended to make to the learning outcomes, which was reflected in their practice in class. They shared these with their class at the start of the lesson. Hazel wrote these up on the whiteboard before the class arrived. Chris talked the class through what he intended the students to learn in each lesson. Fran wrote the learning intentions on the board and shared what she expected to happen in the lesson.

The analysis of the nine lessons taught by the three teachers was made using the integrated framework of Millar (2010) combined with science capabilities. For details of the analysis, see Appendix 6.1 and 6.2. Here we share differences between observed changes in teacher practice with respect to science ideas, procedures, and a scientific approach to investigation; how these changes influenced student learning is discussed in the next section of the chapter. In Phase 1 conceptual knowledge was an intended outcome in all nine lessons, but in Phase 2 science, ideas were taught in the first of the three lessons, which is indicative of a shift in the teachers' focus on developing the science capabilities. Only Fran made the connection to the science idea in the last of the three lessons (Table 6.1).

Table 6.1 Students should develop their knowledge and understanding of the natural world (observational data)

		Hazel			Chris			Fran		
Investigation number		1	2	3	4	5	6	7	8	9
By doing this activity students should develop their knowledge and understanding of the natural world.		✓	✓	✓	✓	✓	✓	✓	✓	✓
		✓			✓			✓		✓

Table 6.2 Students should learn how to use a piece of laboratory equipment or follow a standard practical procedure

		Hazel			Chris			Fran		
Investigation number		1	2	3	4	5	6	7	8	9
By doing this activity, students should learn how to use a piece of laboratory equipment or follow a standard practical procedure.		✓		✓		✓		✓	✓	✓
							✓			

In Phase 1 students were in year nine and were learning to use the basic science equipment and procedures. This is reflected in Table 6.2 where none of the lessons had a focus on procedural knowledge as the students were competent in using the equipment in Phase 2. Only Chris talked about setting up a water bath and drew a picture on the whiteboard about how to set this up. However, he had not shared this as intended learning with the students. This he explained was because he wanted to keep the focus of the lesson on science capabilities but acknowledged that students needed to know what a water bath was and how to set it up.

The focus in all three lessons of each teacher was on helping students to develop an understanding about a scientific approach to investigation and they did this through the lens of the science capabilities. The main difference was that there was an emphasis on Science capability one (CP1), making observation-based inferences in most lessons. One influence of sharing the finding of, in which they had focussed on CP1, moved the teachers' focus to the other capabilities. Since making observations (CP1) were needed to gather evidence (CP2) and critique evidence (CP3), students continued using this capability although the teachers did not emphasise these (see Table 6.3). Chris offered his reasons for this change:

> I *deliberately* did not have *too many* learning intentions this time, but it was *obvious* that these kids can make observations and explanations, and they are applying them, good on them .... I suppose it is bit like not listing all the key competencies as LOs [learning

Table 6.3 By doing this activity, students should develop their understanding of the scientific approach to inquiry

		Hazel			Chris			Fran		
Investigation number		1	2	3	4	5	6	7	8	9
By doing this activity, students should develop their understanding of the scientific approach to inquiry.		✓ CP1	✓ CP1	✓ CP1 CP2	✓		✓ CP3	✓ CP1	✓ CP1 CP3	✓ CP1
		✓ CP1	✓ CP1 CP2	✓ CP1 CP2	✓ CP2 CP3		✓ CP2 CP3	✓ CP1	✓ CP1 CP3	✓ CP1 CP2

Key: CP 1 Gather and interpret data; CP2 Use evidence; CP3 Critique evidence outcomes] for every lesson ... Like our focus could be on the thinking KC [key competency] so we must teach it ... make sure the kids have opportunity and time to *think* but I don't have to tick off participating and contributing KC because the kids are working in groups .... That might be the focus in another lesson where I want them to improve their skills of participating in a group activity and sharpening their ways of contributing to the group .... trying to be *specific* about what *I want them to learn.* For me that is what this change is about. (The italics are used to show teacher emphasis)

Note: The curriculum aims for students to develop five key competencies: thinking; managing self; using language, symbols, and texts; relating to others; participating and contributing.

All three teachers reduced the number of learning intentions for each practical activity/investigation as suggested by Millar (2010). Although their approaches to teaching were different, all three focussed on particular science capabilities. The students were given opportunities to plan investigations, and then to critique their design. They were encouraged to think about the evidence they needed to collect and to critique the robustness of this evidence. It appeared that teachers were supporting students to understand that science investigation is not necessarily linear and sequential, where they follow a predetermined number of steps to arrive at a known answer.

Student Learning Through Investigation
We next look at the outcomes of the classroom activities in terms of students' learning. One student focus group interview was conducted at the end of each observed lesson. Students also completed a brief reflective survey at the end of the third observed lesson in each class (Appendix 6.3). Documentary evidence in the form of student work and their end-of-topic assessment marks were also collected. We drew upon these to find out what students were actually learning.

Transcripts of class discussions from both phases were analysed using the capabilities framework. There was an increase in student responses about using and critiquing evidence and interpreting representations. The numbers of comments pertaining to observation and inference also increased. It seems that students continued to make observations and write inferences even though it was no longer the teaching focus (see Fig. 6.9), perhaps an indication of their applying the capability they had learnt the previous year. As the observation data included teacher–student discourse, there was a teacher-led emphasis on critiquing evidence.

There is a similarity in what was observed in the classroom (Fig. 6.9) and what students talked about in focus groups (Fig. 6.10) which indicates that teachers' learning intentions regarding the capabilities were being met. Observations continued to be made and students used them as evidence for their inferences.

There is additional evidence that students were developing some ideas about using and critiquing evidence. For example, in Fran's class students carried out a metal carbonate and acid reaction. The students tested the gas produced to find out if it was oxygen, hydrogen, or carbon dioxide, rather than just testing for CO_2 and confirming that it was carbon dioxide. 'Miss says you can't just test for CO_2, be sure that there is no oxygen or hydrogen present'. When probed during class this student said, 'Well, we mixed baking soda with acid. Baking soda has carbon,

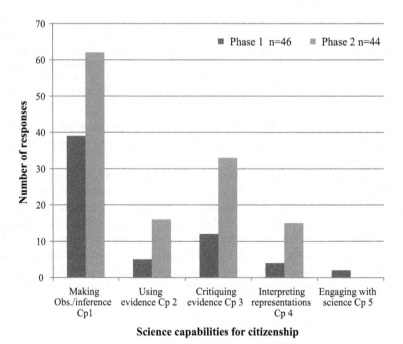

Fig. 6.9 Students' capability related responses from classroom observation

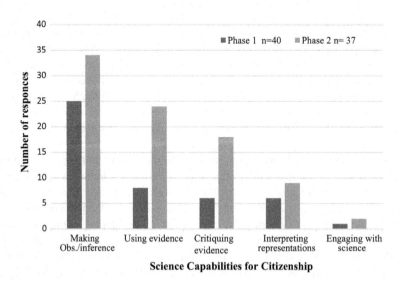

Fig. 6.10 Students' science capability related responses from focus groups in Phase 1 and 2

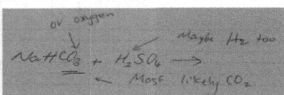

Fig. 6.11 Acid-carbonate reaction and student explaining the need to check both positive and negative results

hydrogen, and oxygen and the acid has hydrogen. It is most likely to be CO_2, but it is better to check to see if there was any oxygen or hydrogen.' He illustrated (Fig. 6.11):

Students in Hazel's class were learning how to carry out a fair testing type of investigation and Hazel had taken the opportunity to teach them about setting up tables and the use of units. As discussed earlier, the students appeared to understand why outliers needed to be addressed and they demonstrated some understanding of this idea. In Phase 2 the students were able to identify patterns in their observation; however, this was not evident in Phase 1. Similarly, Appendix 6.2, Table 6.7 shows students having a better understanding of science ideas in but class observations show little evidence of this (see the last row of Table 6.4). Is this because it was not an intended learning outcome? Did teacher focus on capabilities and not emphasising relevant science ideas make a difference? These questions remain unanswered.

Further analysis of observation data showed that the students were developing some understanding of particular aspects of science investigation. Table 6.5 shows in Phase 1 'selecting a question to investigate' and 'planning' were limited to one lesson but this was common for most lessons in Phase 2. There was also evidence of students making evidence-based conclusions, critiquing evidence, and communicating their findings (see key to the codes used at the end of the table).

Students completed a questionnaire at the end of the third lesson (Appendix 6.3). The data from 61 questionnaires (84% response rate) from the three classes ($n = 72$) were analysed. Figure 6.12 shows what students said they did in the lesson that was similar to what scientists do. Their written responses are congruent with the observations made in class, and the focus group feedback. We are not able to compare this in the two phases as we only used the questionnaire at the end of Phase 2.

In the questionnaire, students were also asked what they thought they had learnt in each of these lessons. All responses to this question were almost exclusively about substantive science ideas, for example, 'We learnt about magnesium and HCl

Table 6.4 Student learning through investigation (from classroom observations)

Teachers		Hazel			Chris			Fran		
Investigation number		1	2	3	4	5	6	7	8	9
Students can recall an observable feature of an object, or material, or event.		✓		✓	✓	✓	✓	✓	✓	✓
		✓			✓			✓		✓
Students can recall a 'pattern' in observations (e.g., a similarity (s), difference (d), trend (t), relationship (r).)			✓		✓p ✓s/d ✓t ✓r		✓	✓	✓	✓
			✓					✓		
Students have a better understanding of a scientific idea (c), or explanation (e), or model (m), or theory (t).		✓ c m	✓ c m		✓ c m	✓ c m	✓ c	✓ c/e m	✓ t m	✓ c/e m

Table 6.5 Students' observed understandings of aspects of science investigation

In this table Hazel is T1, Chris is T2, and Fran is T3

Teachers		Hazel			Chris			Fran		
Investigation number		1 T1	2 T1	3 T1	4 T2	5 T2	6 T2	7 T3	8 T3	9 T3
Students have a better *general understanding* of science investigation					✓			✓		
		✓			✓			✓	✓	
Students have a better understanding of some specific aspect of scientific enquiry		c, f	f	c, d	a, b, c, e, f, g	f	e, f	h	f, i	d, f
		a, b, c, f	a, b, d	c, b, d,	a, b, g, i	d, e, i	g, h	a, b, e, f	a, b, d, g	e, f, h, i

Key:

a   How to identify a good investigation
b   How to plan a strategy for collecting data to address a question
c   How to choose equipment for an investigation
d   How to present data clearly
Note: Codes h and i have been added to the original framework

e   How to analyse data to reveal or display patterns
f   How to draw and present conclusions based on evidence
g   How to assess how confident you can be that a conclusion is correct
h   How to critique evidence
i   How to communicate findings

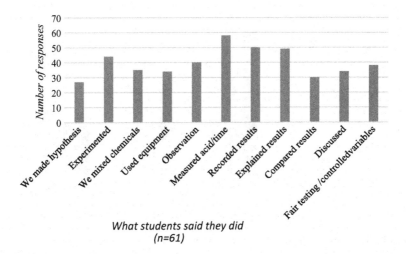

What students said they did
(n=61)

Fig. 6.12 Students' views about similarities between their investigation and scientists' investigations

reaction'; 'physical and chemical change'; 'how higher temperature makes the moliqules [as spelt by the student] move faster'. Is it that the students see science learning as knowing the canonical science ideas? Even though there is evidence of their developing procedural and NOS understandings, students themselves seem to see the ideas about science as part of *doing* science rather than *learning* science.

Students were assessed through a pen and paper test on the fair testing type of investigation (sourced from the Assessment Resource Bank, see https://arbs.nzcer. org.nz/). The test had a focus on thinking with evidence. It was marked using criteria and provided four possible grades: Not achieved, Achieved (describe), Merit (explain) and Excellence (discuss). The test took place approximately three weeks after the observed lessons. It was marked by the teachers, checked by a researcher, and the results are presented in Fig. 6.13. Less than 10% of the participants did not reach an achieved grade.

Overall, we can say that practical work was effective at level one of the framework (Abrahams & Millar, 2008) introduced earlier, that is [brief summary]: The students were able to do with objects and ideas what their teachers intended them to do and were able to think and talk about the science ideas that they had been learning. There is evidence that the students were able to demonstrate an understanding of the ideas about the nature of science that the teachers intended them to (level 2 of the framework) during the lesson, in the focus group, and in the questionnaire, they completed at the end of the set of lessons. The observation data also show that some students retained these ideas from one lesson to the next. However, we only have the results of the end-of-topic test held three weeks after the observed lessons that show that 90% had an Achieved or better grade. This evidence we understand is weak as we do not have a pretest to compare it with.

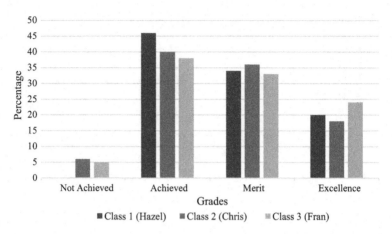

Fig. 6.13 End of topic test for all three classes in

In relation to the science capabilities, which are a current focus in New Zealand, we could say that across the two years students did improve in their ability to make observations and offer evidence-based explanations. Some were able to not only critique evidence but also to critique their method. These findings go some way towards showing that when teachers are cognisant of the need to have fewer intended outcomes from an investigation there are positive learning outcomes for the students. The study does not report on progression of these ideas or the learning demands of the tasks that students engaged in and what progression in the capabilities might look like. We can say that in the observed lessons students were learning from and about science investigation, which is a useful finding—although they themselves did not articulate their learning in this way.

6.4   Summary

The intent of the *Teacher Lead Research Initiative* (as described in Chap. 1) is collaborative in nature and design, and we did not want to *tell* the teachers *what to do*. Our support was in helping teachers to implement the enhancements they wanted to make and collect research evidence of the influence of these changes in practice on student learning. What can we learn from the evidence we collected about secondary teachers making changes to refine and enhance their practice in order to improve student learning from science investigation? First, Hazel, Chris and Fran engaged with the research literature to identify aspects they wanted to improve—they used the support provided.

Regarding the changes made and their impact, evidence suggests that teachers were very clear about what they intended the students to learn from a particular investigation and shared this with the students. Millar's (2010) framework was useful to the teachers for planning to enhance the effectiveness of student learning from science investigation. It was also a useful framework for analysing the classroom observations and focus group interviews.

Overall, this study has given us evidence of student learning from the investigations they engaged in, although we cannot be certain that the learning was retained over the long term. But clear purpose (Hodson, 1990, 2014), not too many intended outcomes (Abrahams & Millar, 2008; Millar, 2010), and emphasis on teaching about the nature of science did lead to the intended learning outcomes for investigating in science. The pedagogical approaches used by the teachers appear to have supported student learning. Abrahams and Reiss (2012) suggest that reviewing each practical task or a reflection at the end of the lesson, as was common extant practice for these teachers, helps consolidate the learning. It appears to have been useful for learning in these classes. The teachers wanted the students to develop their science capabilities; to that end, it appears that the students were at least using some of the capabilities taught. The students were being encouraged to critique the evidence and make evidence-based conclusions, which is promoted by Osborne's (2014) notion of developing investigative understanding as practiced.

Consider these

1. What changes could teachers make to their practice that would better develop science capabilities useful for citizenship?
2. When planning for science investigation how can a balance be made between substantive ideas, the other two purposes for learning: developing investigative skills and understanding about the nature of science?

Appendix 6.1

A checklist for analysing and comparing up to 10 practical activities
   1 Learning objective (s) (or intended learning outcome(s))

	Activity number	1	2	3	4	5	6	7	8	9	10
	1.1 Objective **(in general terms)**(Enter '1' for **main objective**; '2' if necessary for a **subsidiary objective**)										
A	By doing this activity students should develop their knowledge and understanding of the natural world.										
B	By doing this activity, students should learn how to use a piece of laboratory equipment or follow a standard practical procedure										
C	By doing this activity, students should develop their understanding of the scientific approach to enquiry how to use										
A1	Students can recall an observable feature of an object, or material, or event										
A2	Students can recall a 'pattern' in observations (e.g. a similarity, difference, trend, relationship)										
A3 B1	Students have a better understanding of a scientific idea, or explanation, or model, idea, or theory Students can use a piece of equipment, or follow a practical procedure, that they have not previously met										
B2 C1	Students are better at using a piece of equipment, or follow a practical procedure, that they have not previously met Students have a better *general understanding* of enquiry										
C2	Students have a better understanding of some specific aspect of scientific enquiry										

For C2, ↓ rather than simply ticking √ box, enter letters to indicate the specific aspects being taught as follows:

a How to identify a good investigation	e How to analyse data to reveal or display patterns
b How to plan a strategy for collecting data to address a question	f How to draw and present conclusions based on evidence
c How to choose equipment for an investigation	g How to assess how confident you can be that a conclusion is correct
d How to present data clearly	*h How to critique evidence i. communicating (Additions to the original framework)*

Millar (2010). Analysing practical science activities to assess and improve their effectiveness. The Association of Science Education. Retrieved from
http://www.york.ac.uk/media/educationalstudies/documents/research/Analysing%20practical%20activities.pdf

Appendix 6.2

This framework by Millar (2010) was used to compare the change in practice a framework suggested by Millar (2010). It allows a comparison of up to 10 lessons. We used this framework to analyse the nine observed lessons in Phase 1 with the 9 lessons in Phase 2. We have also analysed and added the science capabilities for citizenship in Tables one and two (Appendix 6.2). In this table, Hazel is T1, Chris is T2 and Fran is T3. Science capabilities are identified as CP1 Gather and interpret data; CP2 Use evidence; and CP3 Critique evidence. To make sense of the details in the last two rows of the table see the key below the table (Tables 6.6 and 6.7).

Appendix 6.3: Student Reflective Survey

What did you learn?   Name

What did you observe or measure during the practical?	What did you infer from the observations or measurements?

1   Summarise what you learnt in this lesson about the topic you are studying:

2   a. What did you do today that is similar to what scientists do?

      b. Is this something new you have learnt?

THANK YOU

Table 6.6 Analysis of nine investigations in the three observed classes (Phase 1)

Activity number		1 (T1)	2 (T1)	3 (T1)	4 (T2)	5 (T2)	6 (T2)	7 (T3)	8 (T3)	9 T3)	10
1.1 Objective **(in general terms)**(Enter '1' for **main objective**; '2' if necessary for a **subsidiary**											
A	By doing this activity students should develop their knowledge and understanding of the natural world.	✓	✓	✓	✓	✓	✓	✓	✓	✓	
B	By doing this activity, students should learn how to use a piece of laboratory equipment or follow a standard practical procedure.	✓		✓		✓		✓	✓	✓	
C	By doing this activity, students should develop their understanding of the scientific approach to inquiry.	✓CP1	✓ CP1	✓ CP1 CP2	✓		✓ CP3	✓ CP1	✓ CP1 CP3	✓ CP1	
1.2 **Learning objectives (more specifically)** (Tick ✓ one box in each group for which you have entered a number above)											
A1	Students can recall an observable feature of an object, or material, or event.	✓		✓	✓	✓	✓	✓	✓	✓	
A2	Students can recall a 'pattern' in observations (e.g., a similarity (s), difference (d), trend (t), relationship (r).		✓		✓p ✓s/d ✓t ✓r		✓	✓	✓	✓	
A3	Students have a better understanding of a scientific idea (c), or explanation (e), or model (m), or theory (t).	✓ c m	✓ c m		✓ c m	✓ c m	✓ c	✓ c/e m	✓ t m	✓ c/e m	
B1	Students can use a piece of equipment, or follow a practical procedure that they have not previously met.	✓	✓		✓	✓		✓	✓	✓	
B2	Students are better at using a piece of equipment, or following a practical procedure, that they have not previously met.	✓	✓	✓	✓	✓		✓	✓	✓	
C1	Students have a better *general understanding* of inquiry.				✓			✓			
C2	Students have a better understanding of some specific aspect of scientific inquiry.	c, f	f	c, d	a, b, c, e, f, g	f	e, f	h	f, i	d, f	

For C2, rather than simply ticking √ box, enter letters to indicate the specific aspects being taught as follows:

a How to identify a good investigation

b How to plan a strategy for collecting data to address a question

c How to choose equipment for an investigation

d How to present data clearly

e How to analyse data to reveal or display patterns

f How to draw and present conclusions based on evidence

g How to assess how confident you can be that a conclusion is correct

h How to critique evidence

i . communicating

Table 6.7 Analysis of nine investigations in the three observed classes (Phase 2)

	Activity number	1 (T1)	2 (T1)	3 (T1)	4 (T2)	5 (T2)	6 (T2)	7 (T3)	8 (T3)	9 (T3)
A	By doing this activity students should develop their knowledge and understanding of the natural world.	✓			✓			✓		✓
B	By doing this activity, students should learn how to use a piece of laboratory equipment or follow a standard practical procedure									
C	By doing this activity, students should develop their understanding of the scientific approach to enquiry	✓CP1	✓ CP1 CP2	✓ CP1 CP2	✓ CP2 CP3		✓ CP2 CP3	✓ CP1	✓ CP1 CP3	✓ CP1 CP2
A1	Students can recall an observable feature of an object, or material, or event	✓			✓			✓		✓
A2	Students can recall a 'pattern' in observations (e.g., a similarity (s), difference (d), trend (t), relationship (r).		✓					✓		
A3	Students have a better understanding of a scientific idea (c), or explanation (e), or model (m) , or theory (t)									
B1	Students can use a piece of equipment, or follow a practical procedure, that they have not previously met	✓			✓			✓	✓	✓
B2	Students are better at using a piece of equipment, or follow a practical procedure, that they have not previously met.	✓			✓			✓		✓
C1	Students have a better *general understanding* of enquiry	✓			✓			✓	✓	
C2	Students have a better understanding of some specific aspect of scientific enquiry	a, b, c, f	a,b,d	c, b, d,	a, b, g, i	d, e, i	g, h	a, b, e, f	a, b, d, g	e, f, h, i

For C2, rather than simply ticking √ box, enter letters to indicate the specific aspects being taught as follows:

a How to identify a good investigation

b How to plan a strategy for collecting data to address a question

c How to choose equipment for an investigation

d How to present data clearly

e How to analyse data to reveal or display patterns

f How to draw and present conclusions based on evidence

g How to assess how confident you can be that a conclusion is correct

h How to critique evidence

i communicating

Key: CP 1 Gather and interpret data; CP2 Use evidence; CP3 Critique evidence

References

Abrahams, I., & Millar, R. (2008). Does practical work really work? A study of the effectiveness of practical work as a teaching and learning method in school science. *International Journal of Science Education, 30*(14), 1945–1969.

Abrahams, I., & Reiss, M. J. (2012). Practical work: Its effectiveness in primary and secondary schools in England. *Journal of Research in Science Teaching, 49*(8), 1035–1055.

Hodson, D. (1990). A critical look at practical work in school science. *School Science Review, 71* (256), 33–40.

Hodson, D. (2014). Learning science, learning about science, doing science: Different goals demand different learning methods. *International Journal of Science Education, 36*(15), 2534–2553.

Millar, R. (2010). *Analysing practical science activities to assess and improve their effectiveness.* Hatfield: Association for Science Education. Retrieved from http://www.york.ac.uk/media/educationalstudies/documents/research/Analysing%20practical%20activities.pdf.

Ministry of Education. (2007). *The New Zealand curriculum.* Wellington: Learning Media.

Osborne, J. (2014). Teaching scientific practices: Meeting the challenge of change. *Journal of Science Teacher Education, 25*(2), 177–196.

Chapter 7
Enhancing Learning Through School Science Investigation

In this chapter, we reflect on our findings and their implications. We consider the nature of student learning that we observed and the teachers' beliefs and practices that supported this learning. We discuss changes in practice and what is needed to support such change. Finally, we review the role of school science investigation and the potential it has for supporting/scaffolding/driving student learning. In thinking about these findings, it needs to be borne in mind that they are case study findings and so the ability to generalise is limited. We have, therefore, provided details about the methodology, setting and participants so that the appropriateness of applying the findings to other schools and contexts may be judged by the reader.

7.1 Teacher Beliefs About Science Investigation

We start with teacher beliefs about science investigation. Each of the five participating teachers identified a central role for science investigation in their science teaching. They saw it as providing rich opportunities not simply for building substantive knowledge, but for developing students' investigative skills and their understanding of how science works. Student ownership of investigation was considered to be the ideal approach; however, particularly, for secondary teachers, this was not always pragmatically possible because of assessment and time constraint. They dealt with this constraint by selecting investigations that fitted best with the substantive ideas that they needed to teach as part of the school curriculum. Primary teachers were less constrained in the selection of contexts for investigation; student interest and engagement were key factors in their choice.

Opportunities for student ownership of at least some aspects of the investigation were important to all teachers when selecting contexts and topics; they recognised multiple purposes for learning from science investigation. They saw some conflict in teaching for substantive understanding versus teaching about the nature of science. Some were committed to students developing the criticality essential for

© Springer Nature Singapore Pte Ltd. 2018
A. Moeed and D. Anderson, *Learning through School Science Investigation*,
https://doi.org/10.1007/978-981-13-1616-6_7

scientific literacy and recognised the value of participation in science investigation as a means to this end.

7.2   Approaches to Science Investigation

Goldsworthy, Watson, and Wood-Robinson (1998) recommend primary students need to experience various types of investigation. We observed this with students experiencing a variety of approaches to the investigation in line with the aim of the curriculum. Exploration was particularly common in the primary classrooms; however, students also had the opportunity to classify and identify, and use models and the older primary students considered elements of fairness when designing their investigations. In secondary classrooms, we observed students carrying out exploration, fair testing, classifying, pattern seeking and using models. We saw no evidence of students making things or developing systems as an investigative approach. Internationally and in New Zealand, research suggests that fair testing is the most common investigation used in primary and secondary schools (Goldsworthy et al., 1998; Hipkins et al., 2002; Hume & Coll, 2008). This may be due to an emphasis on fair testing in national science assessment in secondary school and in commonly used resources at primary level. However, participating teachers demonstrated a broader conceptualisation of science investigation. This may be because of their prior access to, and involvement in professional development supporting the expectations of the curriculum.

While all the participating teachers believed in the importance of student ownership of investigation, we observed very few examples of genuinely open-ended investigations during which students had opportunities to make decisions and solve problems (Roberts, 2009). That said, this rare opportunity was afforded to students in a secondary setting when they were investigating ice in Chris's class. Chris's pedagogy was not constrained by lack of time and assessment pressures; he appeared strongly committed to providing students with authentic experiences, possibly because he has worked as a scientist in the past. At the primary level, degrees of student ownership were observed as teachers scaffolded the students through the investigative process. However, students were given long periods of exploration to follow their own lines of investigation as they played with the equipment provided. In Phase 1 Alison made extensive use of students' wonderings and questions as drivers for investigation; this changed in Phase 2 when she prioritised substantive understandings and investigations were very much teacher selected.

7.3   Teacher Practices that Support Substantive and Syntactic Understanding

We found that all primary and secondary teachers in this research provided opportunities for students to develop substantive ideas, investigative skills, understanding of the nature of science and the Science Capabilities for Citizenship, although the balance between them changed between the phases of the research. To facilitate learning, they organised specific equipment that supported development of substantive ideas and allowed time so that the students could explore and investigate in a focused manner. Alison planned and modelled wondering and supported the students to work with their own wonderings to come up with investigable questions. Both primary teachers used questioning to help students investigate in an orderly manner, encouraged careful observation and rigour, and helped students identify and control the range of variables. Although developing substantive ideas and procedural skills for their own sake was often a focus in Phase 1 for the secondary teachers, there was a shift to integrating skill development as part of a specific investigation in Phase 2. Substantive ideas, while still a focus, were not the main outcomes that teachers were looking for in most investigations.

Both primary and secondary teachers provided opportunities for development of the Science Capabilities for Citizenship, which are seen as a functional way of addressing the Nature of Science strand of the New Zealand curriculum. However, all teachers focused mainly on the 'Gathering and Interpreting Data' capability in Phase 1. At secondary level, there was some evidence of opportunity to develop other capabilities, for example, using evidence to form conclusions. The development of the other capabilities became a focus for Phase 2. In primary schools, the teachers used the opportunities provided through science investigation to make scientific practice and values explicit to students. They connected what the students were doing in their investigations with ways that scientists behave. Overt connections to scientific ways of thinking and behaving were not so explicit in secondary school in Phase 1. In Phase 2, being scientific was talked about more frequently: 'In science we use evidence to back our claims; look at robustness of the data; make accurate measurements' and so forth.

Throughout this book, we have used the three major purposes for learning through science investigation—developing substantive ideas, learning investigative skills and building understanding of the nature of science—as a frame for examining student learning and teacher practice. Our current interpretation of our observations is that all student learning through school science investigation, whether substantive or syntactic (i.e. developing investigative skills and understanding about science), begins with the domain of objects (Millar, 2004; Tiberghien, 2000). This is shown in Fig. 7.1. Science is empirically based and seeks to develop explanations of the physical and natural world, so science investigations by definition all concern the objects that form the world and the observations that we can make about them. The development of theories that explain observations is the goal of science. Tiberghien (2000) calls this theoretical realm the domain of

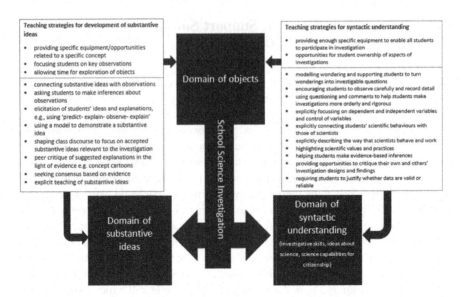

Fig. 7.1 Teacher strategies that support substantive and syntactic understanding. Adapted from Tiberghien (2000) and Millar (2004)

ideas. Millar (2004) suggests that the role of practical work in school science is to help students connect the domain of objects to this domain of ideas.

Our observations of classroom practice lead us to suggest that school science investigation provides further opportunities. We differentiate the domain of ideas into two domains: the *domain of substantive ideas* and the *domain of syntactic understanding*.

We have deliberately used 'understanding' to describe the syntactic aspect, as we noticed that some students were learning not just ideas about how science worked and what constituted scientific practice and behaviour, but were *becoming* scientific in their approach and engaging in scientific practices. They not only knew how to make careful observations but did so in their own investigating, knowing that they should. In other words, they were, as Bull (2015) suggests in her discussion of the dispositional aspect of the science capabilities for citizenship, not only able but also ready and willing to engage with science. To us, this development is more akin to 'learning to be', rather than simply learning an idea. For example, a group of students in Fran's class, as described in Chap. 6, was considering the gas produced from the reaction between sodium bicarbonate and sulphuric acid. They found carbon dioxide was present but went on to test for hydrogen and oxygen to ensure that no other possible gas was also present. They were able to justify their action—just because carbon dioxide is present, does not mean that hydrogen and oxygen are not produced from the reaction as well.

Teachers played a major role in providing and facilitating opportunities for learning in both domains. We have summarised the practices and strategies they used to support learning in each domain in Fig. 7.1.

We see learning in science as underpinned by both constructivist and socio-cultural theory. In developing substantive ideas, students can be seen to be considering their own schema in the light of other possible ideas—the opportunity to observe and discuss possible explanations provides a time for assimilation and integration of new ideas, demonstrating a constructivist approach to learning. The discussion that supports adoption and clarification of these ideas aligns with social constructivism (Anderson, 2007). Learning to participate in scientific ways of doing, thinking and being is much more akin to a sociocultural approach to learning. The teaching strategies that are used for these purposes involve providing opportunities to participate in practices, modelling and making values explicit. These approaches align with sociocultural theories about learning as participation in a community of practice (Rogoff, 2003; Wenger, 1998). Whether students are learning to be scientists, or learning how to engage with science as citizens, they need a lived reality of what it looks like—to learn how to behave and think, not just to know more ideas (Allchin, 2011). We are not saying that the development of substantive ideas is unimportant, simply that the two kinds of learning require different pedagogical approaches, as Hodson (2014) points out. As Fig. 7.1 shows, participation in school science investigation provides opportunities for both kinds of learning. How teachers facilitate and emphasise these two types of learning appears to influence the learning outcomes exhibited by students, as we discuss next.

7.4   The Influence of Change in Practice on Student Learning Through Science Investigation

What teachers know and believe is important to learn is influential in the opportunities they provide for the students and, therefore, has a major impact on student learning. As discussed above, we found that the majority of what students learnt was dependent on where teacher talk was focussed. The primary teachers had been involved in professional development which had highlighted the need for students to develop understandings about the nature of science. We observed that in Phase 1 primary teachers' main focus was on syntactic learning outcomes. For instance, both primary teachers talked about looking closely, observing carefully and persevering, which the students learned to do. In Patsy's class, students recognised these behaviours as important learning. In Phase 2, we noticed a change in the primary teachers' focus. Alison found teaching the abstract substantive ideas associated with bubbles problematic for her young students in Phase 1. In Phase 2, she chose a topic that would be less abstract to her students' everyday lives. They, consequently, spent more time talking about substantive science ideas and this resulted in students developing more substantive ideas. In Patsy's class, although the students were engaged in investigations with a greater degree of student ownership, the focus of teacher talk before and after each investigation was largely on the abstract and complex substantive ideas about heat transfer. Most students thought that learning about heat was the major focus.

In secondary school, in Phase 1 teachers had several learning outcomes for each investigation; teacher talk was focussed on substantive ideas as well as on making careful observations. This was reflected in the evidence we have of student learning. In Phase 2, teachers gave primacy to students developing science capabilities and teacher talk was focussed on looking for evidence, being critical of the evidence collected, and critiquing the design of their investigation, and less on substantive ideas. This was again reflected in what students learnt. Taking on board insights from published research, the secondary teachers had fewer learning outcomes for each investigation and this helped students to have the time to be thoughtful about what they were doing and to work and talk with their peers. Findings were shared by groups, and teachers carefully crafted the discussion to guide the students towards the intended goals. The teachers' skill lay in not accepting the first correct answer or saying what was wrong, but rather in managing to elicit multiple explanations and getting students to discuss and agree on the most evidence-based and plausible ideas. The nature and use of teacher talk in the development of students' thinking and learning was a focus of work by Mortimer and Scott (2003). Our findings show congruence with their ideas.

A practice that supported the opportunities described above was a time for reflection and discussion at the end of the science investigation, which became common in all the primary and secondary classes we observed. Hipkins et al. (2002) suggested that positive learning outcomes are achieved when 'students' ideas are seen to be valued and are used by the teacher to help the students reflect on and move towards the conceptual goals' (p. 114). As the teachers developed more heightened awareness of checking whether the intended learning outcomes were met, the end of investigation reflection became more tightly focussed on the intended learning outcomes.

7.5 Teacher Agency

Sometimes, when teachers participate in classroom research, the intervention is externally imposed. What we mean is that teachers have little input into the research design, and often little choice in deciding what they wish to improve in their practice. The Teaching and Learning Research Initiative (TLRI), which funded this project, encourages collaboration between researchers and teachers. In this research, the teachers had the choice to decide which changes they wanted to make to their practice. They did so based on findings that were shared with them in a non-judgemental way, and by considering published research about effective practice. Together with time and opportunity to reflect, the teachers identified relevant and specific changes to implement, with a focus on achieving better learning outcomes for their students.

Baird (1988) suggests that 'the future of science education does not lie primarily in curriculum or technology but with the teachers of science' (p. 70). This quote highlights the importance of the role of the teacher in implementing changes. Too often changes are made to the curriculum, and hence, expected student outcomes and achievement without providing adequate time, support and resources to enable teachers to enact them effectively. Teachers are also often not party to the thinking behind curriculum decisions. Our study highlights that when teachers have ownership and understand reasons for change in practice, and are provided with the necessary support, the intended curriculum can become the implemented curriculum.

7.6 Science Investigations: Further Opportunities for Learning

We want to finish by refocussing on the role of school science investigation. Our observations have confirmed the central role that science investigations have in providing opportunities for learning in science. Through school science investigation, students have the opportunity to consider and develop the important ideas that form the body of science knowledge, and thus current understanding about the way the world works. School science investigation also provides an opportunity for students to learn to investigate in scientific ways and to learn what it means to be scientific. Recent discussions about the school curriculum for science internationally have highlighted science as practice. For example, in the United States, the change in the education mandated by the framework for K-12 National Research Council (2012) and the Next Generation Science Standards (Achieve, 2012) shows a shift from teaching science as inquiry (investigation) to teaching science as practice (Osborne, 2014). Some suggest that science investigations have, therefore, been superseded by teaching science as practice. The standards cite the following scientific practices:

- Asking questions and defining problems
- Developing and using models
- Planning and carrying out investigations
- Analysing and interpreting data
- Using mathematical and computational thinking, constructing explanations, and designing solutions
- Engaging in argument from evidence obtaining, evaluating and communicating information. (Osborne, 2014, p. 179)

These scientific practices are described in international curricula in different ways and, as evident in Chap. 2, are consistent with the requirements of the Nature of

Science strand of the New Zealand Curriculum. New Zealand's 'Science Capabilities for Citizenship' also include the development of core scientific practices. However, an understanding about scientists' socially negotiated norms and values is more complex and is unlikely to emerge just through engaging in scientific practices. We think it requires a connection being made about why scientists engage in these practices and we saw some evidence of it in the observed classes. Teaching science as practice does not replace the need for school science investigation; in fact, as we have observed, school science investigation provides the opportunity for the learning of science as practice.

Simply engaging in science investigation does not directly result in students learning a substantive idea, or learning how to investigate, or understanding what constitutes science. The role of the teacher is critical in connecting and guiding students towards these different outcomes. Hodson (2014) argues that for effective learning in each of these areas, teachers need to be clear about the intended goal and focus of their teaching and talk accordingly. As Fran concluded:

> … you don't just do practical work or investigation, I think about what I want the students to learn, and then decide what might be the most appropriate way for the students to learn that.

As documented in this study, there is plenty of evidence of students doing investigations in their science classes and evidence that they learn from doing them. This finding is heartening as previous research suggests that, in general, students learn little from engaging in practical work (Abrahams & Millar, 2008; Abrahams & Reiss, 2012). However, as we have been reviewing our findings, it occurs to us that there is potential for school science investigation to achieve broader curriculum objectives beyond the substantive and syntactic learning we have described. The New Zealand curriculum in its entirety allows for and indeed encourages more comprehensive and wide-ranging investigations and outcomes than those we observed. For example, the Investigating in Science strand is situated alongside the Participating and Contributing strand of the Nature of Science expectations for learning. This strand emphasises bringing a scientific perspective to issues-based decision-making and action taking. In addition, values such as ecological sustainability and caring for the environment, integrity and acting ethically, respect for self, others and human rights are to be encouraged, modelled and explored (Ministry of Education, 2007). The curriculum encourages that all learning should make use of the natural connections that exist between learning areas to contribute to a broad general education that builds values and key competencies. The investigations we observed and have described tended to be very science outcome focussed. However, we saw glimpses of the types of investigation opportunities indicated by the broader New Zealand curriculum. For example, during the water topic in Phase 1, students investigated the water quality of the local stream. Alison had created opportunities for her students to engage in ongoing investigations, which included establishing a butterfly enclosure, making decisions about what kinds of plants would attract butterflies and growing and caring for the plants. Opportunities for learning went beyond science and included language development and mathematical learning. The establishment of the butterfly enclosure led to student engagement in science beyond the observed science lessons. Students were seen spontaneously rushing into the classroom to find magnifying glasses so they could take a closer look, pointing out the changes they observed to their parents, being responsible, curious and caring. These incidental observations highlight the potential for engagement, learning and inculcation of values afforded by long-term and broad-ranging investigations.

7.7 Suggestions for Practice

Teachers in our research wanted to improve student learning through making research-based changes to their practice. Our findings are in congruence with many scholars in suggesting what can be done to improve teaching so that it enhances learning.

In sum, the key messages are as follows:

- Be clear about your purpose for each investigation
- Provide equipment and opportunities that relate specifically to the purpose
- Be clear about what the students will do
- Be explicit about the intended learning from the investigation
- Elicit and work with students' ideas
- Explicitly connect what students do with scientific practices and values
- Encourage and guide discussion that considers evidence and
- Determine whether the intended learning has taken place.

References

Abrahams, I., & Millar, R. (2008). Does practical work really work? A study of the effectiveness of practical work as a teaching and learning method in school science. *International Journal of Science Education, 30*(14), 1945–1969.

Abrahams, I., & Reiss, M. J. (2012). Practical work: Its effectiveness in primary and secondary schools in England. *Journal of Research in Science Teaching, 49*(8), 1035–1055.

Allchin, D. (2011). Evaluating knowledge of the nature of (whole) science. *Science Education, 95*(3), 518–542.

Anderson, C. W. (2007). Perspectives on science learning. In S. K. Abell & N. G. Lederman (Eds.), *Handbook of research on science education* (pp. 3–30). Mahweh, NJ: Lawrence Erlbaum.

Baird, J. (1988). Teachers in science education. In P. Fensham (Ed.), *Development and dilemmas in science education*. London: The Farmer Press.

Bull, A. (2015). *Capabilities for living and lifelong learning: What's science got to do with it?*. Wellington, NZ: New Zealand Council for Educational Research.

Goldsworthy, A., Watson, R., & Wood-Robinson, V. (1998). Sometimes it's not fair. *Primary Science Review, 53*, 15–17.

Hipkins, R., Bolstad, R., Baker, R., Jones, A., Barker, M., Bell, B., ... Taylor, I. (2002). *Curriculum, learning and effective pedagogy: A literature review in science education.* Wellington: Ministry of Education.

Hodson, D. (2014). Learning science, learning about science, doing science: Different goals demand different learning methods. *International Journal of Science Education, 36*(15), 2534–2553.

Hume, A., & Coll, R. (2008). Student experiences of carrying out a practical science investigation under direction. *International Journal of Science Education, 30*(9), 1201–1228.

Millar, R. (2004). The role of practical work in the teaching and learning of science. *High school science laboratories: Role and vision.* Retrieved from http://www.informalscience.org/images/research/Robin_Millar_Final_Paper.pdf.

Ministry of Education. (2007). *The New Zealand curriculum.* Wellington: Learning Media.

Mortimer, E. F., & Scott, P. H. (2003). *Meaning making in secondary science classrooms.* Maidenhead: Open University Press.

National Research Council. (2012). *A framework for K-12 science education: Practices, crosscutting concepts, and core ideas.* Washington, DC: Committee on a Conceptual Framework for New K-12 Science Education Standards. Board on Science Education, Division of Behavioral and Social Sciences and Education.

Osborne, J. (2014). Teaching scientific practices: Meeting the challenge of change. *Journal of Science Teacher Education, 25*(2), 177–196.

Roberts, R. (2009). Can teaching about evidence encourage a creative approach in open-ended investigations? *School Science Review, 90,* 31–38.

Rogoff, B. (2003). *The cultural nature of human development.* New York: Oxford University Press.

Tiberghien, A. (2000). Designing teaching situations in the secondary school. In R. Millar, J. Leach, & J. Osborne (Eds.), *Improving science education: The contribution of research* (pp. 27–47). Buckingham, UK: Open University Press.

Wenger, E. (1998). *Communities of practice: Learning, meaning and identity.* Cambridge: Cambridge University Press.

Printed in the United States
By Bookmasters